# NOTES ON ELEMENTARY PARTICLE PHYSICS

# NOTES ON ELEMENTARY PARTICLE PHYSICS

*by*

H. MUIRHEAD

*Department of Physics*
*University of Liverpool*

PERGAMON PRESS
OXFORD NEW YORK TORONTO
SYDNEY BRAUNSCHWEIG

Pergamon Press Ltd., Headington Hill Hall, Oxford
Pergamon Press Inc., Maxwell House, Fairview Park, Elmsford, New York 10523
Pergamon of Canada Ltd., 207 Queen's Quay West, Toronto 1
Pergamon Press (Aust.) Pty. Ltd., 19a Boundary Street, Rushcutters Bay, N.S.W. 2011, Australia
Vieweg & Sohn GmbH, Burgplatz 1, Braunschweig

---



First edition 1972

Library of Congress Catalog Card No. 79-156263

*Printed in Germany*

**08 016550 8**

# CONTENTS

# PREFACE

THE notes in this book are a more or less verbatim account of an introductory course of thirty lectures on elementary particle physics given to first-year graduate students at the University of Liverpool during the winter of 1969/70.

In the lectures an attempt was made to convey the current ideas on the state of the subject to a group of students who had a good grounding in quantum mechanics, but whose knowledge of particle physics was limited to the descriptive level. With this background in mind, I hope that the book may also prove of use to undergraduates who wish to delve a little deeper into the subject than they might normally go in an undergraduate course.

Since it is a lecture course of limited extent some topics have not been dealt with in this book as fully as they might have been in a more formal and comprehensive textbook. I have therefore occasionally referred the student to my larger book *The Physics of Elementary Particles* (Pergamon) — these references appear in the text as P.E.P. However, such references come under the heading of "additional reading matter", and are not essential to the material covered in this book.

No problems are given in the book. The students taking the course were given problems at monthly intervals after it was completed. These were taken mainly from preprints or the letters sections of journals.

In developing material for this book, I have been greatly indebted to Dr. G. R. Allcock of the Department of Theoretical Physics, University of Liverpool for many enlightening discussions. I should also like to thank Mr. P. Grossmann of this Department for assistance in checking the proofs.

Lastly I am indebted to the following authors and editors for permission to reproduce certain figures and tables appearing in the text:

V. Barger;
R. J. N. Phillips;
C. Schmid;
J. Steinberger;
R. R. Wilson;
CERN Information Service;
DESY Information Service;
*Physical Review Letters*;
*Physics Letters* (North Holland Publishing Co.);
*Reviews of Modern Physics*;
SLAC Information Service;
University of California Press.

H. Muirhead

CHAPTER 1

# TERMINOLOGY

## 1.1. The Major Types of Particles

In the early 1930's a list of known elementary particles would have consisted of the electron, proton and neutron. The well-understood electromagnetic interactions between the electrons outside the nucleus and protons inside accounted for the whole of atomic physics. The strong binding of protons and neutrons within the nucleus indicated the existence of forces much stronger than the electromagnetic ones.

Even at this stage, however, the discovery of further particles could reasonably be expected. The work of Dirac (1928, 1929) on a relativistic theory for the electrons had led to the prediction of the existence of what we would nowadays call antiparticles, i.e. particles with properties opposite to those of particles.‡ The first antiparticle, a positively charged electron (positron), was found in the cosmic radiation (Anderson, 1932; Blackett and Occhialini, 1933). Since that time both antiprotons and antineutrons have also been found.

The existence of a further type of particle had also been predicted. Experiments on nuclear $\beta$-decay had led to the uncomfortable result that apparently neither energy nor angular momentum were conserved in this process. Pauli (1933) suggested that this dilemma might be resolved by postulating the existence of a further particle—the neutrino. This particle would have to be massless (or nearly so), have no electrical charge and spin $\frac{1}{2}$ (in units of $\hbar$). A particle without electric charge is difficult to detect at the best of times. The neutrino had the additional awkward

‡ What one describes as particle and what as antiparticle is merely a matter of definition. Since our world consists of electrons and protons we refer to their charge opposites as antiparticles.

property that it interacts only weakly with matter (§ 1.2); consequently its existence was not proved until 23 years after Pauli's prediction (Cowan *et al.*, 1956).

Nowadays the number of elementary particles runs to several hundred and grows each year (see Appendix A1). The choice of name has been unfortunate, but there would appear to be little point in changing it at this stage. Because of their large numbers, classification and terminology is all important and the rest of this chapter will be concerned with introducing the appropriate words.

Despite their large numbers the elementary particles break down to three basic groups, which are determined by the way they interact. They are:

(1) the photon—$\gamma$,
(2) leptons—neutrino $\nu$, electron $e$, muon $\mu$,
(3) hadrons—meson $M$, baryon $B$.

We shall define the muon later. The mesons are systems which decay, and the decay chain ends in leptons or photons, for example the charged kaon $K^+$ can decay as

$$K^+ \to \overset{\mu^+\nu}{\overset{\uparrow}{\pi^+}}\ \underset{\gamma\gamma}{\underset{\downarrow}{\pi^0}}.$$

$B$ indicates baryons which are particles where the decay chain ends in a proton (the only stable baryon)

$$\Xi^- \to \underset{p\pi^-}{\underset{\downarrow}{\Lambda^0}}\ \overset{\mu^-\bar{\nu}}{\overset{\uparrow}{\pi^-}}.$$

The basic groups interact with each other in one or more of three independent ways—in order of increasing strength these are the weak, electromagnetic and strong interactions. As we have already mentioned, the electromagnetic and strong interactions are responsible for atomic and nuclear binding respectively; the weak interactions occur in nuclear $\beta$-decay and in other places which we shall discuss later. The strengths of the interactions will be examined in § 1.2.

The interactions of the particles with each other determine their classification; as examples:

(1) electromagnetic interactions always involve photons which couple to the charge or magnetic moment of a particle;
(2) only particles which can interact strongly are defined as hadrons. At the same time, however, hadrons can interact electromagnetically by virtue of their intrinsic electric charges or magnetic moments.

The complete list of interactions of the photon, leptons and hadrons with each other are given in Table 1.1, where W, E and S denote weak, electromagnetic and strong interactions respectively.

TABLE 1.1.

| | Photon $\gamma$ | Lepton $\nu, e, \mu$ | Hadron $M, B$ |
|---|---|---|---|
| Photon | E | E | E |
| Lepton | E | E, W | E, W |
| Hadron | E | E, W | S, E, W |

The symbol $\mu$ in our lepton group refers to the muon. Although this particle has decay properties similar to those defined for mesons,

$$\mu \rightarrow e\nu\bar{\nu},$$

its interaction properties place it firmly in the lepton camp. We shall discuss the triple entry for hadron–hadron interactions in § 1.3.

## 1.2. Coupling Strengths

The electromagnetic, strong and weak interactions are characterised by their coupling strengths. They are as follows:

### *1. Electromagnetic*

In electromagnetic interactions the photons interact with the charge or magnetic moment of the other particles. In these interactions a term

$$\alpha = \frac{e^2}{4\pi\hbar c} \sim \frac{1}{137}$$

always appears, for example in the Bohr picture of the hydrogen atom we find the following expressions for the properties of the lowest orbit:

$$\text{radius} \qquad = \frac{1}{\alpha m}\,\frac{\hbar}{c}$$

$$\text{velocity of electron in orbit} = \alpha c$$

$$\text{energy} \qquad = -\frac{1}{2} m\alpha^2 c^2.$$

*2. Strong*

The coupling parameter is

$$\frac{g^2}{4\pi\hbar c} \sim 10$$

where $g$ has the same units as $e$.

*3. Weak*

Because of historical reasons the coupling strength for weak interactions was defined with different dimensional properties than those for electromagnetic and strong interactions. For our present purposes a dimensionless number

$$C \sim 10^{-5}$$

will suffice; we shall return to the weak coupling strength in § 2.5.

One can form a rough idea about how these values arise by considering the measured cross-sections $\sigma$ for typical processes

| | $\sigma$ experiment |
|---|---|
| Compton scattering from protons | $\gamma p \rightarrow \gamma p \sim 10^{-32}\ \text{cm}^2$ |
| pion–nucleon scattering | $\pi p \rightarrow \pi p \sim 10^{-26}\ \text{cm}^2$ |
| lepton scattering | $\nu p \rightarrow n\mu^+ \sim 10^{-38}\ \text{cm}^2$. |

Now a cross-section is an area and, therefore, the square of a length. The only length that the processes have in common is the nucleon

Compton wavelength‡

$$r = \frac{\hbar}{Mc} = \frac{1}{M} \quad (\hbar = c = 1)$$

$$\sim 10^{-14} \text{ cm.}$$

If we make the educated guess that

$$\text{transition probability} \equiv \sigma \sim \left| \frac{\text{Strength function}}{M} \right|^2$$

$$\text{transition amplitude} = \sqrt{\sigma} \sim \frac{\text{Strength function}}{M}$$

then for

$$\gamma p \to \gamma p \quad \text{strength function} \sim 10^{-2},$$

$$\pi p \to \pi p \quad \sim 10,$$

$$\nu p \to n\mu^+ \quad \sim 10^{-5}.$$

Now consider decay processes; the important thing measured here is the lifetime. Since the total transition probability per unit time $\Gamma$ is the inverse of the lifetime $\tau$ we may guess

$$\frac{1}{\tau} = \Gamma \propto |\text{Strength function}|^2$$

hence

$$\tau_{\text{weak}}/\tau_{\text{strong}} \sim 10^{12}.$$

Experimentally lifetimes for strong decays can be measured by the uncertainty principle and give

$$\tau_{\text{strong}} \sim 10^{-22} \text{ sec}$$

hence we would expect

$$\tau_{\text{weak}} \sim 10^{-22} \times 10^{12} \sim 10^{-10} \text{ sec.}$$

This sort of time is about right for weak decays providing no inhibiting factors occur. Hyperons and kaons (§ 1.3) have lifetimes in this region.

‡ It is convenient to work with units such that $\hbar = c = 1$ in particle physics, since these terms constantly occur. The appropriate rules for dimensional transformation are discussed in Appendix A2.

It should be noted that mixtures of the basic interactions can occur, for example the photoproduction of mesons $\gamma p \to p\pi^0$. This process has a cross-section $\sim 10^{-2}$ times smaller than that for pion–nucleon scattering.

In addition to the three basic interactions one other is known in nature—that of gravitation. It is much weaker than the weak interaction and is outside the scope of this book. Speculations have also been made that superweak and superstrong interactions exist in elementary particle physics—their existence has yet to be proved.

## 1.3. Strange Particles

The name strange particle arose from historical reasons. Mesons (kaons) and baryons (hyperons) were found in the cosmic radiation‡ with a frequency of occurrence which implied that they were produced in strong interactions, yet they decayed by processes of the type

$$K^+ \to \pi^+\pi^0 \qquad \Lambda^0 \to p\pi^-$$

with lifetimes characteristic of weak interactions. This was surprising since the decay products were strongly interacting particles. Hence, we have the triple entry in our table of § 1.1 for hadron–hadron interactions. Particles with the double property of production by strong interactions and decay into hadrons by weak interactions are called strange particles. We shall return to their properties in the next section. Examples of strange particles are the kaon, $K$, and hyperons $\Lambda$, $\Sigma$, $\Xi$, $\Omega$.

## 1.4. The Gell-Mann, Nishijima Scheme

### 1.4(a). Isospin and hadrons

During the 1930's studies in nuclear structure physics led to the concept that the strong nuclear forces were charge independent, that is if a neutron is replaced by a proton in any system then, providing the particles are in the same state of angular momentum, the energy of the

‡ A detailed description of the history of elementary particle physics is given in Chapter 1 of *P.E.P.*

system is unaltered. This point is well illustrated in the energy level structure shown in Fig. 1.1 for the nuclei $B^{12}$, $C^{12}$ and $N^{12}$. It can be seen that many levels correspond closely to each other in the three nuclei.

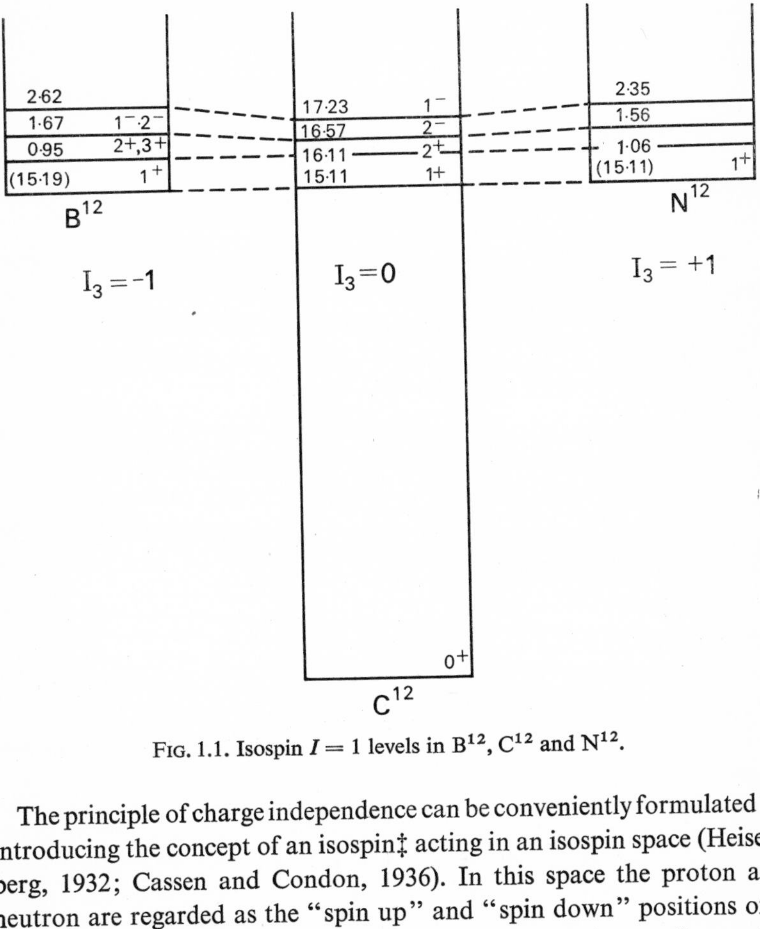


FIG. 1.1. Isospin $I = 1$ levels in $B^{12}$, $C^{12}$ and $N^{12}$.

The principle of charge independence can be conveniently formulated by introducing the concept of an isospin‡ acting in an isospin space (Heisenberg, 1932; Cassen and Condon, 1936). In this space the proton and neutron are regarded as the "spin up" and "spin down" positions of a state called the nucleon with isospin $\frac{1}{2}$. Spin up and down refers to the 3-axis in this space.

‡ Isospin is also called isotopic spin and isobaric spin.

For our present purposes we shall define an isospin $I$ through the requirement

$$\text{number of charged states of a hadron} = 2I + 1. \qquad (1.4.1)$$

Thus we can write

$$\text{nucleon } N = p, n \qquad I = \tfrac{1}{2},$$

$$\text{pion } \pi = \pi^+, \pi^0, \pi^- \quad I = 1.$$

We further define the $I_3$ values through the relation

$$\frac{Q}{e} = I_3 + \frac{B}{2} \qquad (1.4.2)$$

where $Q$ is the charge on the particle and $B$ is the baryon number (+1 for one baryon, −1 for one antibaryon, zero for mesons). Thus we have for nucleons and pions

| | | | | | |
|---|---|---|---|---|---|
| $I_3$ | $-1$ | $-\frac{1}{2}$ | $0$ | $+\frac{1}{2}$ | $+1$ |
| $N$ | | $n$ | | $p$ | |
| $\pi$ | $\pi^-$ | | $\pi^0$ | | $\pi^+$ |

Similarly the equivalent levels in our nuclear triplet $B^{12}$, $C^{12}$ and $N^{12}$ can be regarded as the $I_3 = -1, 0, +1$ components of an $I = 1$ state.

Since $I_3$ is linearly proportional to charge, the charge independence of strong interactions implies that they are independent of $I_3$. If we assume that $I$ and $I_3$ behave in isospin space like the spin parameters $j$ and $j_3$ in ordinary space (i.e. like vectors and scalars respectively), then we should expect to find their values conserved. This is indeed found for strong interactions

| | $\pi^-$ | $p \rightarrow$ | $\pi^0$ | $n$ |
|---|---|---|---|---|
| **I** | 1 | $\frac{1}{2}$ | 1 | $\frac{1}{2}$ |
| $I_3$ | $-1$ | $\frac{1}{2}$ | 0 | $-\frac{1}{2}$. |

Now consider

| | $d$ | $d \rightarrow$ | $He^4$ | $\pi^0$ |
|---|---|---|---|---|
| **I** | 0 | 0 | 0 | 1. |

Isospin is not conserved and we would expect the process not to occur. Experimental searches have indicated that the cross-section is at least 100 times smaller than that expected if isospin conservation was ignored (Poirier and Pripstein, 1963).

Now consider electromagnetic interactions. Since the photon links itself to electrical charge they are manifestly charge dependent,‡ but $I_3$ is conserved since charge is conserved. Since, for example, we observe the radiative decay of the $\Sigma^0$ hyperon (the photon is not a hadron and so it does not have isospin quantum numbers)

$$\begin{array}{ccc} & \Sigma^0 \rightarrow & \Lambda^0\gamma \\ \mathbf{I} & 1 & 0 \\ I_3 & 0 & 0 \end{array}$$

we conclude that $I_3$ holds but **I** conservation is broken in electromagnetic interactions.

In purely hadronic weak interactions both **I** and $I_3$ conservation is broken, however

$$\begin{array}{cccc} & \Lambda^0 \rightarrow & p & \pi^- \\ \mathbf{I} & 0 & \frac{1}{2} & 1 \\ I_3 & 0 & +\frac{1}{2} & -1. \end{array}$$

Two properties which are conserved in this interaction are charge and baryon number (as in all weak interactions)

$$\begin{array}{cccc} & \Lambda^0 \rightarrow & p & \pi^- \\ Q & 0 & +e & -e \\ B & 1 & 1 & 0. \end{array}$$

### 1.4(b). The strangeness quantum number

The study of elementary particles has been helped enormously by the emergence of simple conservation laws which are found to hold or break

‡ For example, the electromagnetic potential between two electrons or two positrons is $e^2/r$, but is $-e^2/r$ between an electron and a positron ($r$ is the distance of separation).

in a well-defined way. The examples we quoted in the previous section are each typical of their class of interaction.

Consider the question of strange particles. Their production in strong interactions and decay by weak interactions led Pais (1952) to speculate that perhaps some property was being conserved in strong interactions and violated in weak. He, therefore, suggested that perhaps strange particles were produced in pairs and that these pairs had opposite properties so that the overall algebraic effect was zero. When they decayed, they did so on their own and the property was violated. This suggestion coincided with the introduction of a 4-GeV accelerator in the U.S.A. where the following sequence of processes was observed in a diffusion chamber

$$\begin{array}{ccc} & p\pi^- & \\ & \uparrow & \\ \pi^- p \rightarrow & \Lambda^0 & K^0. \\ & & \downarrow \\ & & \pi^+\pi^- \end{array}$$

The observation of events such as these led Gell-Mann (1953, 1956) and Nishijima (1955) to an independent classification of the properties of the hadrons. This is summarised in the relation

$$\frac{Q}{e} = I_3 + \frac{1}{2}(B + S) \qquad (1.4.3)$$

where $S$ is the strangeness quantim number. Coupled with this relation went the rule—strong interactions conserve **I**, $I_3$, $S$; purely hadronic weak interactions violate **I**, $I_3$ and $S$‡

| | $Q$ | $B$ | **I** | $I_3$ | $S$ |
|---|---|---|---|---|---|
| strong interactions | ✓ | ✓ | ✓ | ✓ | ✓ |
| weak interactions | ✓ | ✓ | × | × | × |

An examination of the observed examples of strong and purely hadronic weak interactions led to the assignments given in Table 1.2 for the quantum

‡ We have included conservation of charge $Q$ and baryon number $B$ in this list. No examples of the violation of these two properties have ever been found in weak, electromagnetic or strong interactions, for example

$$\pi^+ p \nleftrightarrow \pi^0 p$$
$$\pi^+ p \nleftrightarrow \pi^+\pi^+ p\bar{p}.$$

numbers of the particles which are stable or decay into hadrons by weak interactions.

At the time of the first assignment the $\Xi^0$ and $\Sigma^0$ hyperons had not been found, but were included in the table to provide a self-consistent picture. Their existence was later confirmed. The $\Omega^-$ hyperon is a later addition (§ 7.2 (d)).

TABLE 1.2.

| $I$ | $I_3$ | | | | | $B$ | $S$ |
|---|---|---|---|---|---|---|---|
| | $+1$ | $+\frac{1}{2}$ | $0$ | $-\frac{1}{2}$ | $-1$ | | |
| 0 | | | $\Omega^-$ | | | 1 | $-3$ |
| $\frac{1}{2}$ | | $\Xi^0$ | | $\Xi^-$ | | 1 | $-2$ |
| 1 | $\Sigma^+$ | | $\Sigma^0$ | | $\Sigma^-$ | 1 | $-1$ |
| 0 | | | $\Lambda^0$ | | | 1 | $-1$ |
| $\frac{1}{2}$ | | $p$ | | $n$ | | 1 | 0 |
| $\frac{1}{2}$ | | $K^+$ | | $K^0$ | | 0 | $+1$ |
| $\frac{1}{2}$ | | $\bar{K}^0$ | | $K^-$ | | 0 | $-1$ |
| 1 | $\pi^+$ | | $\pi^0$ | | $\pi^-$ | 0 | 0 |

## 1.5. Experimental Techniques

The discovery and elucidation of the properties of the particles is bound up with the energy and intensity of the primary beams of particles.

Most of the early work in elementary particle physics was done using the cosmic radiation as the primary beam. This provides a continuous energy spectrum of electrons and protons. The intensity of the beam is always weak (total flux of cosmic rays at sea level $\sim 10^{-2}$ particles per square centimetre per second) and falls rapidly with increasing energy; nevertheless it is still the only source of ultra high energy particles.

The large accelerators provide beams of protons and electrons at controllable energies with fluxes of $\sim 10^{12}$ particles per second. The machines with energies greater then 4 GeV (1 GeV $= 10^9$ eV) are given in Table 1.3.

Most of the institutions listed are joint projects, since the facilities are so expensive, for example CERN (Centre for European Nuclear Research) is a consortium of twelve European countries who contribute in pro-

TABLE 1.3.

PROTON SYNCHROTRONS

| Laboratory and location | Maximum energy in GeV |
|---|---|
| Serpukhov, U.S.S.R. | 76 |
| BNL, U.S.A. | 30 |
| CERN, Switzerland | 25 |
| Argonne, U.S.A. | 10 |
| Dubna, U.S.S.R. | 10 |
| Berkeley, U.S.A. | 8 |
| Rutherford, U.K. | 7 |
| Saclay, France | 4 |

ELECTRON ACCELERATORS

| Laboratory and location | Maximum energy in GeV |
|---|---|
| Stanford linac, U.S.A. | 20 |
| Cornell synchrotron, U.S.A. | 10 |
| Cambridge synchrotron U.S.A. | 6 |
| DESY synchrotron, Germany | 6 |
| NINA synchrotron, U.K. | 4 |

portion to their net national incomes. The proton synchrotrons cause the acceleration of protons in an orbit of constant radius $R$ given by

$$pc = BeR$$

where $p$ is the momentum of the protons, $c$ the velocity of light and $B$ the strength of the magnetic field. During the acceleration cycle $B$ is continuously increased and at the same time so also is the frequency $f$ of the accelerating voltage in order to keep $R$ constant. This frequency is given by

$$f = \frac{1}{2\pi R} \frac{p}{E}$$

where $E$ is the energy of the particle. In the electron synchrotrons only $B$ is increased, since by virtue of their low mass the electrons effectively reach the velocity of light at low energies ($\sim$1 MeV) and thereafter the frequency $f$ can be kept constant. The electrons and protons are normally injected into the synchrotrons from linacs producing particles with energies of $\sim 10$ MeV.

The linac‡ accelerates particles by driving them down a straight tube (wave guide) where they are subjected to pulses of R.F. power. The Stanford linac is 2 miles in length.

In addition to the accelerators listed in Table 1.3 a 200-GeV proton synchrotron (extendable to 400 GeV) is under construction at Batavia, U.S.A., and a 300-GeV machine is being planned in Europe. A project of great current interest is the intersecting storage rings (ISR) at CERN. This involves taking two 25-GeV proton beams from the CERN proton synchrotron and allowing them to collide head on. The 50 GeV of energy then available in each collision is equivalent to a proton beam of energy 1300 GeV striking a hydrogen target [this result can easily be shown with the aid of equation (A.5.7)]. However, the probability of such collisions occurring is much smaller than in a conventional machine. In addition to the ISR project, successful colliding beam machines have been built for work with electron accelerators. Many interesting experiments have already been done with them (see Chapter 5).

The beams of particles from the accelerators are normally used to strike targets to produce secondary beams and these are transported to the appropriate experimental areas around the machines. The problems of efficient beam transport both inside the accelerators and in the secondary beams are very sophisticated and require enormous amounts of work for effective operation. A typical beam layout is shown in Fig. 1.2.

Finally the beam strikes a target and the results of the interaction must be detected. Most detectors rely on the principle of ionisation—the displacement of electrons in an atom by the high energy particle. The resultant physical processes are then exploited and lead to effects on a macroscopic scale. The effects occur in many ways:

1. In the bubble chamber a liquid (usually hydrogen) is kept near its boiling point and the passage of a particle leaves a trail of bubbles which can be photographed.

‡ An abbreviation for linear accelerator.

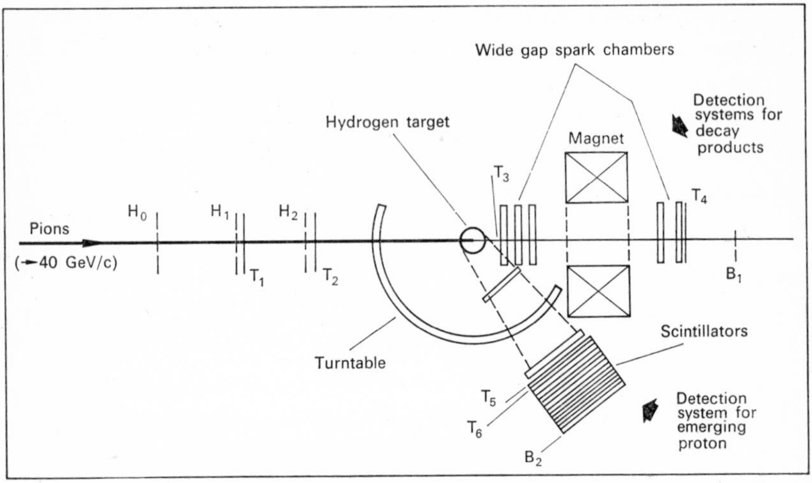


FIG. 1.2. Schematic representation of the CERN-Serpukhov collaboration experiment to detect the mass spectra of mesons ($X^-$) in the process $\pi^-p \rightarrow pX^-$ at 40 GeV/c (CERN, 1970). Counter hodoscopes $H_0$, $H_1$ and $H_2$ monitor the incoming beam. The counter sequence $T_1$ to $T_6$ must all record simultaneously whilst $B_1$ and $B_2$ must not register for a good event (the firing of $B_1$ implies that the incoming pion passed through the apparatus, whilst it is kinematically impossible for a proton to reach $B_2$). If a favourable sequence ($T_1$ to $T_6$ on, $B_1$ and $B_2$ off) occurs, the electronics switch on the spark chambers to obtain details of $X^-$ and the event is recorded.

2. The ionisation can cause scintillations in certain chemical compounds; these are detected by photomultipliers which convert the light signal to electrical pulses which are passed to electronic circuits. The apparatus is known as a scintillation counter.

3. Spark and discharge chambers. These are in many ways a cross between the previous two techniques. Basically they comprise a pair of metal plates or wire meshes with a rare gas between them and also a difference in potential. The passage of a fast particle nearly perpendicularly to the plates causes a spark or discharge depending on the magnitude of the voltage. These effects can then be detected

electronically (or optically in the case of the spark). The great advantage of the technique is that it is possible to locate exactly where the particle passed between the pair of plates or meshes whilst at the same time producing an electronic signal. An assembly of many plates or meshes is normally used.

Each technique has its own advantages. The bubble chamber is normally both target and detector and provides a complete record (with $4\pi$ geometry) of all the charged particles participating in a reaction. At the same time it is unselective and gathers information on both wanted and unwanted reactions. The electronic techniques are more selective but normally are incapable of providing the complete coverage of a reaction which the bubble chamber offers.

Normally in all reactions the kinematic properties of the ingoing and outgoing particles are measured in order to provide dynamic information on the related processes. The momenta of the ingoing and outgoing particles are normally determined by the application of magnetic fields

$$pc = BeR.$$

Thus a measurement of the radius of curvature $R$ gives the momentum $p$. For certain reactions this information, together with the angles of emission of the outgoing particles, is adequate. In other situations the velocity of the particle $\beta$ (in units of velocity of light $c = 1$) must also be known in order to establish its mass and hence its identity

$$p = \frac{m\beta}{\sqrt{1 - \beta^2}}.$$

The velocity can be measured by determining the ionisation which varies roughly as $1/\beta^2$. This information can be obtained from the number of bubbles per unit length in the bubble chamber, or from the electronic pulse heights in the scintillation counter and discharge chamber techniques. One further approach is possible when $\beta \to 1$; this is the Cerenkov counter which relies on the principle that if a particle travels faster than light in a medium of refractive index $\mu$, then the light (Cerenkov radiation) emitted is concentrated at an angle $\vartheta$ given by

$$\cos\vartheta = \frac{1}{\beta\mu}.$$

This technique is of great importance at very high energies.

Finally at virtually every stage from planning to final analysis an enormous computing effort is required for the successful prosecution of an experiment in elementary particle physics.

CHAPTER 2

# TRANSITION AMPLITUDES AND PROBABILITIES

## 2.1. Notation

As stated in Chapter 1 we shall normally work in units where $\hbar = c = 1$ (see Appendix A2). The Minkowski convention will be used for four-vectors—this implies imaginary fourth components. Thus we shall write

$$\text{space coordinates } x_\lambda \equiv x_1, x_2, x_3, x_4 \equiv x, y, z, ict \equiv \mathbf{x}, it,$$

$$\text{four momenta } p_\lambda \equiv p_1, p_2, p_3, p_4 \equiv p_x, p_y, p_z, \frac{iE}{c} \equiv \mathbf{p}, iE,$$

$$k_\lambda \equiv k_1, k_2, k_3, k_4 \equiv k_x, k_y, k_z, \frac{i\omega}{c} \equiv \mathbf{k}, i\omega,$$

$$\text{curent density } j_\lambda \equiv j_1, j_2, j_3, j_4 \equiv j_x, j_y, j_z, ic\varrho \equiv \mathbf{j}, i\varrho.$$

Whenever possible the symbol $p$ will be kept for fermions and $k$ for bosons.

In general, we shall use the convention that the subscripts for four-vectors are denoted by Greek indices and three vectors by Latin indices. We shall also drop summation symbols when considering repeated suffices; in addition when summations involve the momentum four-vector we shall occasionally drop the indices altogether in order to simplify notation

$$\begin{aligned} \sum_\lambda p_\lambda p_\lambda &= p_\lambda p_\lambda = p_\lambda^2 = p^2 = \mathbf{p}^2 - E^2 = -m^2 \\ \sum_\lambda k_\lambda x_\lambda &= k_\lambda x_\lambda = kx = \mathbf{k} \cdot \mathbf{x} - \omega t. \end{aligned} \tag{2.1.1}$$

Note, however, that on occasion the momentum three-vector will be represented by

$$|\mathbf{p}| = p. \tag{2.1.2}$$

Finally the Dirac notation of kets ($|\alpha\rangle$) and bras ($\langle\alpha|$) will be used to denote quantum mechanical states (Appendix A3).

## 2.2. The Transition Amplitude

### 2.2(a). The S matrix

Heisenberg (1943) introduced an operator — the $S$ operator — whose function was to provide a complete description of a reaction. Supposing we start with an initial state, which we denote by the ket $|\,i\rangle$, then $S\,|\,i\rangle$ yields a complete set of all possible final states. Further, if we are interested in a specific final state $|\,f\rangle$ then the matrix element

$$S_{fi} = \langle f|\, S\, |i\rangle \tag{2.2.1}$$

represents the transition amplitude to $|\,f\rangle$. Thus the transition probability is

$$P_{fi} = |S_{fi}|^2. \tag{2.2.2}$$

If our initial state is normalised

$$\langle i\,|\,i\rangle = 1$$

then $S$ must be a unitary operator. This follows from the completeness relation (A.3.9) for $|\,f\rangle$, and the physical requirement that the sum of probabilities over all final states must be one

$$\sum_f P_{fi} = 1 = \sum_f S_{fi}^* S_{fi} = \sum_f \langle i|\, S^\dagger\, |f\rangle \langle f|\, S\, |i\rangle$$
$$= \sum_f \langle i|\, S^\dagger S\, |i\rangle. \tag{2.2.3}$$

This equation is satisfied if $S$ is unitary‡

$$S^\dagger S = \hat{1} \qquad S^\dagger = S^{-1}. \tag{2.2.4}$$

‡ The symbol $\hat{1}$ represents the unit operator.

The unitarity condition plays an important role in the theory of the $S$ matrix.

## 2.2(b). Invariance principles and the transition amplitude

As we have seen in Chapter 1 conservation laws are of great importance in elementary particle physics. In this section we shall show that the laws are associated with the requirement that the appropriate operators commute with $S$. Let us firstly consider the formal manipulations.

Consider the existence of a unitary operator

$$U^{-1} = U^{\dagger} \quad \text{then} \quad U^{\dagger}U = U^{-1}U = \hat{1}$$

which transforms our states $|i\rangle$ and $|f\rangle$ as

$$|i'\rangle = U\,|i\rangle \qquad |f'\rangle = U\,|f\rangle. \tag{2.2.5}$$

If $\langle f|\,S\,|\,i\rangle$ remains invariant under this process, we can write

$$\langle f|\,S\,|i\rangle = \langle f'|\,S\,|i'\rangle = \langle f|\,U^{-1}SU\,|i\rangle$$

thus

$$S = U^{-1}SU \quad \text{or} \quad [U, S] = 0. \tag{2.2.6}$$

Now let us consider what this means in practice. Supposing an experiment is performed at CERN and then the apparatus is taken to Brookhaven. We should expect the same physical results, that is the experiment should be independent of its displacement. Now we can build the total displacement up from a series of elemental displacements; the appropriate operator for performing this task will be derived in § 3.3. It is

$$U = \hat{1} - ia_{\lambda}P_{\lambda} = \hat{1} - iaP \tag{2.2.7}$$

where $a_{\lambda}$ is the elemental displacement and $P_{\lambda}$ is the four momentum operator

$$P\,|i\rangle = p_i\,|i\rangle \qquad P\,|f\rangle = p_f\,|f\rangle$$

where $p_i$ and $p_f$ represent the total four momenta in initial and final states respectively. Since we expect the physics to be independent of both finite and infinitesimal displacements, equations (2.2.6) and (2.2.7) yield the result

$$[P, S] = 0$$

hence

$$\begin{aligned}\langle f|\,[P, S]\,|i\rangle &= \langle f|\,PS - SP\,|i\rangle \\ &= (p_f - p_i)\,\langle f|\,S\,|i\rangle = 0.\end{aligned} \tag{2.2.8}$$

Thus $p_f = p_i$ and energy and momentum are conserved in the experiment if it is independent of displacements. In a similar fashion invariance under rotations (§ 3.3) implies conservation of angular momentum, and invariance under gauge transformations leads to conservation laws for charge, strangeness and baryon number (a discussion of gauge transformations can be found in § 3.4 (f. 2) and § 5.2 of *P.E.P.*).

### 2.2(c). Additional forms of the transition amplitude

For practical calculations it is often convenient to use a series of reduced matrix operators rather than $S$ alone. They are the $R$, $M$ and $T$ matrices. The reactance matrix $R$ is defined as

$$S = \hat{1} + iR \tag{2.2.9}$$

so that

$$S_{fi} = \langle f|\,\hat{1}\,|i\rangle + i\,\langle f|R|i\rangle = \delta_{fi} + iR_{fi}. \tag{2.2.10}$$

Thus $\delta_{fi} = 1$ implies that "nothing happens".

In the previous section it was shown that the invariance of the $S$ matrix under displacements implied that energy and momentum were conserved between initial and final states. It is often convenient to display the conservation law explicitly by defining a new matrix element $M_{fi}$

$$S_{fi} = \delta_{fi} + i(2\pi)^4\,\delta(p_f - p_i)\,M_{fi} \tag{2.2.11}$$

when the term $(2\pi)^4$ has been extracted for later convenience. The $\delta$-function is a product of four $\delta$-functions, which we can write as

$$\delta(p_f - p_i) = \delta(\mathbf{p}_f - \mathbf{p}_i)\,\delta(E_f - E_i). \tag{2.2.12}$$

Our final matrix element is $T_{fi}$; this is defined as

$$S_{fi} = \delta_{fi} + i(2\pi)^4\,\delta(p_f - p_i)\,\frac{1}{N}\,T_{fi} \tag{2.2.13}$$

where $N$ is a normalisation factor. It is a product of factors $\sqrt{2\omega V}$ for bosons, $\sqrt{VE/m}$ for fermions‡ (one for each particle present), where $\omega$, $E$ are energies and $V$ is a volume of a large "box" which encloses the particles. This box is a convenient fiction which we shall discard later. As we shall see later $T_{fi}$ has the attractive property of Lorentz invariance.

## 2.3. Transition Probabilities

It is apparent from our definitions (2.2.2) and (2.2.10) that the transition probability is given by

$$P_{fi} = |S_{fi}|^2 = (2\pi)^8 \, [\delta(p_f - p_i)]^2 \, |M_{fi}|^2 \tag{2.3.1}$$

since $\delta_{fi} = 0$ for $f \neq i$. If we use the definition of the $\delta$-function (Appendix A4), the double term can be written as

$$\delta(p_f - p_i) \lim_{\substack{V\to\infty \\ T\to\infty}} \frac{1}{(2\pi)^4} \int_{VT} d^4x \, e^{i(p_f - p_i)x} = \delta(p_f - p_i) \frac{VT}{(2\pi)^4}$$

Hence

$$P_{fi} = (2\pi)^4 \, \delta(p_f - p_i) \, VT \, |M_{fi}|^2.$$

Now this represents a transition probability over all space-time. In practice measurements are carried out for a finite time in a restricted volume of space. A more useful definition is the transition probability per unit volume of space-time, which we can write as

$$\omega = \frac{1}{VT} P_{fi} = (2\pi)^4 \, \delta(p_f - p_i) \, |M_{fi}|^2. \tag{2.3.2}$$

In practice two situations are encountered experimentally—the measurement of a cross-section or a decay distribution (which includes lifetimes). We shall treat them separately in the following sections, but before doing so we must introduce an important factor which is common to both situations—phase space.

‡ Bosons obey Bose–Einstein statistics and possess integral spin, whilst fermions satisfy Fermi–Dirac statistics and have half integral spin.

### 2.3(a). Phase space

A single particle in a definite state of motion, i.e. a specified position $x$, $y$, $z$ and momentum $p_x$, $p_y$, $p_z$ can be represented by a point in a six-dimensional position and momentum space. This space is called a phase space.

In classical mechanics no limits can be put on the density of points representing particles in this space. In quantum mechanics the limit is set by the uncertainty principle, where we have the one-dimensional relationship

$$\Delta p \Delta x \geqq (2\pi\hbar).$$

The volume of an elementary cell in a quantum mechanical system is therefore $(2\pi\hbar)^3$, and the total number of states available to a single particle is given by the total volume of phase space divided by $(2\pi\hbar)^3$

$$n = \frac{1}{(2\pi\hbar)^3} \int dx\, dy\, dz\, dp_x\, dp_y\, dp_z = \frac{V}{(2\pi\hbar)^3} \int d\mathbf{p}.$$

In practice we are often interested in the number of states available in an element of momentum space and so we write

$$dn = \frac{V}{(2\pi\hbar)^3} d\mathbf{p}.$$

If we have $q$ particles present then this relation becomes

$$dN_q = \frac{V^q}{(2\pi\hbar)^{3q}} \prod_{j=1}^{q} d\mathbf{p}_j.$$

### 2.3(b). Cross-sections

We define the cross-section $\sigma$ for the production of $q$ particles by the relation

$$\sigma = \sum_f \frac{\omega}{J} dN_{fq} \tag{2.3.3}$$

where $\omega$ is the transition probability per unit element of space-time (2.3.2), $dN_q$ the phase space weighting for $q$ particles, $J$ is the incident flux density and the summation is over all final states, $f$, compatible with

the initial state. We define $J$ in the following manner. We must have two particles to make a collision, hence in a box of volume $V$ their density per unit volume is $1/V^2$; since these particles move with relative velocity $v$ the effective density (the flux density) becomes

$$J = \frac{v}{V^2}.$$

Thus the expression for the cross-section becomes

$$\sigma = \frac{(2\pi)^4}{v} V^2 \sum_f |M_{fi}|^2 \frac{V^q}{(2\pi)^{3q}} \prod_{j=1}^{q} d\mathbf{p}_{fj}\, \delta(p_f - p_i)$$

where we have set $\hbar = 1$. One final modification must be made to this formula. If the initial particles possess spin but are unpolarised, an averaging over initial spin states must be made; we indicate this requirement by the symbol $\overline{\sum_i}$ and our cross-section becomes

$$\sigma = \frac{(2\pi)^4}{v} V^2 \overline{\sum_i} \sum_f |M_{fi}|^2 \frac{V^q}{(2\pi)^{3q}} \prod_{j=1}^{q} d\mathbf{p}_{fj}\, \delta(p_f - p_i). \qquad (2.3.4)$$

At this stage we can set $V = 1$ for all practical purposes, since $|M_{fi}|^2$ contains $V$ terms which cancel with the $V^{q+2}$ displayed above [§ 4.1 (b)]. The cancellation also occurs explicitly if we substitute the $T$ matrix (2.2.13) for $M_{fi}$, then the cross-section becomes

$$\sigma = \frac{(2\pi)^4}{v} \frac{1}{N_i^2} \overline{\sum_i} \sum_f |T_{fi}|^2 \frac{1}{(2\pi)^{3q}} \prod_{j=1}^{q} \frac{d\mathbf{p}_{fj}}{N_j^2} \delta(p_f - p_i) \qquad (2.3.5)$$

where the individual terms $N^2$ are

$$N^2 = 2\omega \text{ for bosons}$$

$$= E/m \text{ for fermions} \qquad (2.3.6)$$

and the phase space factors $d\mathbf{p}_{fi}/N_j^2$ are Lorentz invariants (Appendix A5). Since both equations (2.3.4) and (2.3.5) must lead to the same results their choice is largely dictated by circumstances; as we shall see later the Lorentz invariant formulae for phase space are easier to handle, but $T_{fi}$ can be more complicated than $M_{fi}$.

### 2.3(c). Decay probabilities

In this situation we are interested in the transition probability per unit time $\Gamma$ and its inverse $\tau$, the mean lifetime of a decaying state. Using our basic expression for $\omega$ (2.3.2) we find

$$\Gamma = \frac{1}{\tau} = \overline{\sum_i} \sum_f V\omega \, dN_{fq}$$

$$= V(2\pi)^4 \overline{\sum_i} \sum_f |M_{fi}|^2 \frac{V^q}{(2\pi)^{3q}} \prod_{j=1}^{q} d\mathbf{p}_{fj} \, \delta(p_f - p_i) \tag{2.3.7}$$

$$= \frac{1}{N_i^2} (2\pi)^4 \overline{\sum_i} \sum_f |T_{fi}|^2 \frac{1}{(2\pi)^{3q}} \prod_{j=1}^{q} \frac{d\mathbf{p}_{fj}}{N_j^2} \delta(p_f - p_i). \tag{2.3.8}$$

### 2.3(d). The relative velocity

In the above expressions for cross-sections and decay probabilities, two kinematic factors appear—the relative velocity $v$ of the incoming particles and the phase space term.

Let us consider the former. In practice two situations arise, firstly the laboratory reference frame with a stationary target particle, in this case $v$ is the velocity of the particles in the incident beam which strikes the target. More often, however, one works in the centre of momentum or $c$-system. This is a reference frame in which the two initial particles (labelled $a$, $b$) have overall momentum zero

$$\mathbf{p}_a + \mathbf{p}_b = 0. \tag{2.3.9}$$

We shall therefore write

$$|\mathbf{p}_a| = |\mathbf{p}_b| = p_c^i. \tag{2.3.10}$$

Thus the relative velocity in the $c$-system is given by

$$v_{ab} = \left| \frac{\mathbf{p}_a}{E_a} - \frac{\mathbf{p}_b}{E_b} \right| = p_c^i \left( \frac{1}{E_a} + \frac{1}{E_b} \right) = p_c^i \frac{E_c}{E_a E_b} \tag{2.3.11}$$

where $E_c$ represents the total energy in the $c$-system, $E_c = E_a + E_b$.

**2.3(e). The phase space factor**

We now consider the phase space factor

$$N_q = \sum_f \frac{1}{(2\pi)^{3q}} \prod_{j=1}^{q} d\mathbf{p}_{fj}\, \delta(p_f - p_i).$$

For our present purposes we may drop the $(2\pi)^3$ and replace the summation by an integral. For a two-particle system we therefore have

$$N_2 = \int d\mathbf{p}_1\, d\mathbf{p}_2\, \delta(\mathbf{p}_1 + \mathbf{p}_2 - \mathbf{p}_i)\, \delta(E_2 + E_2 - E_i). \qquad (2.3.12)$$

For simplicity of writing we shall represent the three vector $d\mathbf{p}$ in spherical coordinates ($d\Omega$ is an element of solid angle)

$$d\mathbf{p} = p^2\, dp\, d\Omega. \qquad (2.3.13)$$

We shall also use the relation

$$E^2 = \mathbf{p}^2 + m^2 \qquad E\, dE = p\, dp$$

where $p = |\mathbf{p}|$. Thus equation (2.3.12) becomes

$$N_2 = \int d\Omega_1\, dp_1\, p_1^2\, \delta[E_1 + \sqrt{m_2^2 + (\mathbf{p}_i - \mathbf{p}_1)^2} - E_i]$$

$$= \int d\Omega_1\, dE_1\, E_1\, p_1\, \delta[E_1 + \sqrt{m_2^2 + (\mathbf{p}_i - \mathbf{p}_1)^2} - E_i].$$

We now make use of the property of the $\delta$-function (Appendix A4)

$$\int dx\, \delta[f(x)] = \left|\frac{\partial f}{\partial x}\right|^{-1}_{x=x_0} \qquad (2.3.14)$$

where $x_0$ is a solution of the equation $f(x) = 0$. In the present situation

$$\left|\frac{\partial f}{\partial E_1}\right| = \left|\frac{\partial}{\partial E_1}(E_1 + E_2 - E_i)\right| = 1 + \frac{\partial E_2}{\partial E_1}$$

$$E_2^2 = m_2^2 + (\mathbf{p}_i - \mathbf{p}_1)^2 = m_2^2 + p_i^2 - 2\mathbf{p}_i \cdot \mathbf{p}_1 + E_1^2 - m_1^2$$

$$2E_2 \frac{\partial E_2}{\partial E_1} = -2\mathbf{p}_i \cdot \mathbf{p}_1 \frac{E_1}{\mathbf{p}_1^2} + 2E_1$$

$$\left|\frac{\partial f}{\partial E_1}\right| = \frac{\mathbf{p}_1^2 E_i - \mathbf{p}_i \cdot \mathbf{p}_1 E_1}{\mathbf{p}_1^2 E_2},$$

hence

$$N_2 = \int d\Omega_1\, E_1 p_1 \left| \frac{\partial f}{\partial E_1} \right|^{-1}$$

$$= \int d\Omega_1 \frac{E_1 E_2 |\mathbf{p}_1|^3}{\mathbf{p}_1^2 E_i - \mathbf{p}_i \cdot \mathbf{p}_1 E_1}. \tag{2.3.15}$$

If we work in the $c$-system we may write

$$\mathbf{p}_i = 0 \qquad E_i = E_c$$

$$|\mathbf{p}_1| = |\mathbf{p}_2| = p_c^f$$

$$N_2 = \int d\Omega_1 \frac{E_1 E_2}{E_c} p_c^f. \tag{2.3.16}$$

This result illustrates one of the advantages of working in the $c$-reference frame, most expressions tend to be much simpler in it.

The approach given above can be extended to three particles; the result (see *P.E.P.*, p. 275, for a derivation) is

$$N_3 = \int d\mathbf{p}_1\, d\Omega_2 \frac{E_2 E_3\, |\mathbf{p}_2|^3}{\mathbf{p}_2^2(E_i - E_1) - (\mathbf{p}_i - \mathbf{p}_1)\cdot \mathbf{p}_2 E_2}. \tag{2.3.17}$$

Both integrations have been stopped when the $\delta$-functions have been exhausted. Since the $\sum_f$ in our expressions for cross-sections and decay rates cover the matrix elements as well as phase space any further integration has to take the functional behaviour of the matrix element into account.

If the integral for three bodies is performed over $d\Omega_2$ rather than $dE_2$ an alternative expression arises

$$N_3 = 2\pi \int d\Omega_1\, dE_1\, dE_2\, E_1 E_2 E_3. \tag{2.3.18}$$

In general, however, three-body phase space integrals (and higher numbers than three) can be handled more easily if they are put in a Lorentz invariant form (Appendix A5). Consider a phase space integral in this form for three particles in their $c$-system

$$N_3' = \int \frac{d\mathbf{p}_1}{E_1} \frac{d\mathbf{p}_2}{E_2} \frac{d\mathbf{p}_3}{E_3} \delta(\mathbf{p}_f - \mathbf{p}_i)\, \delta(E_f - E_i)$$

$$= \int \frac{d\Omega_1\, dp_1\, p_1^2\, d\Omega_2\, dp_2\, p_2^2}{E_1 E_2 E_3} \delta(E_f - E_i)$$

$$= 2\pi \int d\Omega_1\, dE_1\, dE_2 \frac{p_1 p_2}{E_3} d(\cos\vartheta_{12})\, \delta(E_f - E_i).$$

Now

$$\mathbf{p}_3^2 = \mathbf{p}_1^2 + 2\mathbf{p}_1 \cdot \mathbf{p}_2 + \mathbf{p}_2^2$$

$$p_3\, dp_3 = E_3\, dE_3 = p_1 p_2\, d(\cos\vartheta_{12}),$$

hence

$$N_3' = 2\pi \int d\Omega_1\, dE_1\, dE_2\, dE_3\, \delta[(E_1 + E_2 + E_3) - E_i]$$

$$= 2\pi \int d\Omega_1\, dE_1\, dE_2. \tag{2.3.19}$$

This is obviously the same expression as (2.3.18), if the latter is divided by $E_1E_2E_3$; similarly the Lorentz invariant form of two-body phase space is given by equation (2.3.16) with the $E_1E_2$ factor removed

$$N_2' = \int d\Omega_1 \frac{p_c^f}{E_c}. \tag{2.3.20}$$

The Lorentz invariance property of $N'$ allows one to write an ingenious recursion relation (Srivastava and Sudarshan, 1958) relating phase spaces for $q$ and $q-1$ particles. We shall discuss this in Appendix A5.

## 2.4. Cross-sections for Two-body Final States

Consider some general process involving four particles $a$, $b$, $d$, $e$

$$ab \to de.$$

Then the cross-section [equation (2.3.4)] can be written as

$$\sigma = \frac{(2\pi)^4}{v_{ab}} V^2 \overline{\sum_i} \sum_f |M_{fi}|^2 \frac{V^2}{(2\pi)^6} d\mathbf{p}_d\, d\mathbf{p}_e\, \delta(p_f - p_i). \tag{2.4.1}$$

If we work in the $c$-system, then

$$E_a + E_b = E_c = E_d + E_e$$

and equations (2.3.11) and (2.3.16) for $v_{ab}$ and phase space respectively give us the following result for the differential cross per unit solid angle (steradian)

$$\frac{d\sigma}{d\Omega} = \frac{V^4}{(2\pi)^2} \frac{E_a E_b E_d E_e}{E_c^2} \frac{p_c^f}{p_c^i} \overline{\sum_i} \sum_f |M_{fi}|^2 \tag{2.4.2}$$

where the summation over $f$ now refers to spin states and any other property compatible with the initial state since the phase space summation has been made. The removal of the normalisation factors gives us the alternative expression in terms of $T_{fi}$ (2.2.13). A further advantage of the $T_{fi}$ formulation can now be seen—simplicity

$$\frac{d\sigma}{d\Omega} = \frac{1}{(2\pi E_c)^2} \frac{1}{n^2} \frac{p_c^f}{p_c^i} \overline{\sum_i} \sum_f |T_{fi}|^2 \tag{2.4.3}$$

where the prescription for $n^2$ follows from (2.3.6)

$$n^2 = \begin{cases} 16 & \text{four bosons,} \\ \dfrac{4}{m_1 m_2} & \text{two bosons, two fermions,} \\ \dfrac{1}{m_a m_b m_d m_e} & \text{four fermions.} \end{cases} \tag{2.4.4}$$

Furthermore, if $a$ and/or $b$ have spin, then

$$\overline{\sum_i} = \frac{1}{(2s_a + 1)(2s_b + 1)} \sum_i$$

where $s_a$ and $s_b$ are the spins of $a$ and $b$ respectively.

An example of the use of the above equations, without even requiring any detailed information about the nature of $T_{fi}$, was in the determination of the spin of the pion. Cross-sections were determined for the reactions

$$pp \to \pi^+ d \quad \text{(i)}$$

$$\pi^+ d \to pp \quad \text{(ii)}.$$

The experiments were performed at the same $E_c$ value and so by the principle of detailed balance one expects the same transition rate in both directions

$$\sum_i \sum_f |T_{fi}|^2 = \sum_f \sum_i |T_{if}|^2 \tag{2.4.5}$$

where right- and left-hand sides of the equation refer to reactions (i) and (ii) respectively. The averaging over spin states tells us that

$$\overline{\sum_i} = \frac{1}{(2s_p+1)^2} \sum_i \qquad \text{for (i)}$$

$$\overline{\sum_i} = \frac{1}{(2s_\pi+1)(2s_d+1)} \sum_i \qquad \text{for (ii).}$$

Thus by taking the ratio of the measured cross-sections and comparing with equation (2.4.3), one is left with a ratio of momenta and spin weighting factors. The former are known from the kinematics of the reaction, and only $s_\pi$ was left as an unknown. The comparison yielded $2s_\pi + 1 = 1{\cdot}0 \pm 0{\cdot}1$ (Cohen, Crowe and Dumond, 1957), thus $s_\pi = 0$.

## 2.5. Applications of the Phase Space Integral

Some of the nicest applications of the phase space integral occur in weak interactions where many transition rates are determined by a single coupling strength and a phase space term. Consider the $\beta$-decay of the neutron

$$n \to pe^-\bar{\nu}.$$

Equation (2.3.7) then gives us

$$\Gamma_n = \frac{1}{\tau_n} = (2\pi)^4 \overline{\sum_i} \sum_f |M_{fi}|^2 \frac{1}{(2\pi)^9} d\mathbf{p}_p\, d\mathbf{p}_e\, d\mathbf{p}_\nu\, \delta(p_f - p_i) \tag{2.5.1}$$

where $\tau_n$ is the neutron mean lifetime and we have set $V = 1$ since it always cancels.

Now the mass difference of neutron and proton is

$$m_n - m_p = \Delta = 1{\cdot}3 \text{ MeV}.$$

Thus the final particles will have maximum momenta of about 1 MeV and with a comparable maximum energy for the neutrino and electron.

If one considers the relevant expression for phase space (2.3.17) in appropriate notation

$$N_3 = \int d\mathbf{p}_e \, d\Omega_\nu \frac{E_\nu E_p \, |\mathbf{p}_\nu|^3}{\mathbf{p}_\nu^2(m_n - E_e) - \mathbf{p}_e \cdot \mathbf{p}_\nu E_\nu}$$

then since $E_p \sim m_n \sim 1000$ MeV, we can drop the last term in the denominator and make the approximation

$$N_3 = \int d\mathbf{p}_e \, d\Omega_\nu E_\nu p_\nu$$

and so

$$\Gamma_n = \frac{1}{\tau_n} = \frac{1}{(2\pi)^5} \overline{\sum_i} \sum_f |M_{fi}|^2 \, d\mathbf{p}_e \, d\Omega_\nu E_\nu p_\nu . \tag{2.5.2}$$

For later convenience we shall write

$$\overline{\sum_i} \sum_f |M_{fi}|^2 = G^2 \tag{2.5.3}$$

which we shall assume to be a constant; then if we assume the neutrino mass is zero so that $E_\nu = p_\nu$ we find

$$\frac{1}{\tau_n} = \frac{G^2}{(2\pi)^5} \int 4\pi \, d\mathbf{p}_e E_\nu p_\nu$$

$$= \frac{G^2}{(2\pi)^5} (4\pi)^2 \int_{m_e}^{\Delta} dE_e \, E_e p_e (\Delta - E_e)^2 . \tag{2.5.4}$$

If we write $E_e = \eta\Delta$ the above equation becomes

$$\frac{1}{\tau_n} = \frac{G^2}{(2\pi)^5} (4\pi)^2 \Delta^5 \int_{m_e/\Delta}^{1} d\eta \eta \sqrt{\eta^2 - \frac{m_e^2}{\Delta^2}} (1 - \eta)^2 .$$

This equation can be easily evaluated in the limit $m_e/\Delta \to 0$, yielding

$$\Gamma_n = \frac{1}{\tau_n} = \frac{G^2}{(2\pi)^5} (4\pi)^2 \Delta^5 \int_0^1 d\eta \eta^2 (1 - \eta)^2$$

$$= \frac{G^2}{(2\pi)^5} (4\pi)^2 \frac{\Delta^5}{30} = \frac{G^2}{60\pi^3} \Delta^5 . \tag{2.5.5}$$

An exact evaluation gives a blocking factor of 0·47 in front of the above expression, but (2.5.5) is adequate for our purposes. It is apparent from this expression that $G$ cannot be dimensionless. Let us introduce a scaling parameter $\tau_0$ for lifetimes in terms of the nucleon mass $M$

$$\tau_0 = \frac{1}{M} \equiv \left(\frac{\hbar}{Mc^2}\right) \sim 10^{-24} \text{ sec}$$

then since the measured value of $\tau_n$ is about $10^3$ sec

$$\frac{\tau_0}{\tau_n} \sim 10^{-27} = \frac{G^2}{60\pi^3} \frac{\Delta^5}{M} = \frac{(GM^2)^2}{60\pi^3}\left(\frac{\Delta}{M}\right)^5. \tag{2.5.6}$$

Since $\Delta \sim 1{\cdot}3$ MeV, $M \sim 1000$ MeV we therefore find

$$GM^2 \sim 10^{-5} = C \tag{2.5.7}$$

where $C$ is the dimensionless parameter for weak interactions discussed in § 1.2.

Now let us apply our result to muon decay, $\mu \to e\nu\bar{\nu}$. A comparison with equation (2.5.5) suggests an expression

$$\frac{1}{\tau_\mu} \sim \frac{G^2}{60\pi^3} \Delta_\mu^5$$

where

$$\Delta_\mu = m_\mu - m_e \sim 100 \text{ MeV},$$

thus

$$\tau_\mu = \left(\frac{\Delta}{\Delta_\mu}\right)^5 \tau_n \sim 10^{-6} \text{ sec}. \tag{2.5.8}$$

This is satisfactorily close to the measured value. In fact a careful evaluation of muon lifetime using the coupling constants from nuclear $\beta$-decay show a discrepancy with experiment of only 4 per cent; this is believed to arise from the effects of $SU_3$ [§ 6.2(d)].

The same basic expression may be used to evaluate the branching ratio of the pion in the mode $\pi^+ \to \pi^0 e^+ \nu$. The total decay rate is

$$\Gamma_\pi = \frac{1}{\tau_\pi} = \Gamma_{\pi\to\mu\nu} + \Gamma_{\pi\to e\nu} + \Gamma_{\pi\to\pi^0 e\nu}$$

where the measured value of $\tau_\pi \sim 10^{-8}$ sec. Equation (2.5.5) in the present situation yields

$$\Gamma_{\pi\to\pi^0 e\nu} = \frac{G^2}{60\pi^3}\Delta_\pi^5 \sim 1 \text{ sec}^{-1}$$

where

$$\Delta_\pi = m_{\pi+} - m_{\pi^0} \sim 4{\cdot}5 \text{ MeV},$$

hence the expected branching ratio is

$$\frac{\Gamma_{\pi\to\pi^0 e\nu}}{\Gamma_\pi} \sim 10^{-8}$$

which agrees nicely with experiment (Depommier *et al.*, 1963).

Next consider a neutrino induced reaction

$$\bar{\nu}p \to nl^+$$

where $l^+$ represents a positively charged lepton (electron or muon). If we adapt equation (2.4.2) to our present requirements, the cross-section will be given by

$$\frac{d\sigma}{d\Omega} = \frac{1}{(2\pi)^2}\frac{E_\nu E_n E_p E_l}{E_c^2}\frac{p_c^f}{p_c^i}\overline{\sum_i}\sum_f |M_{fi}|^2. \tag{2.5.9}$$

Now for moderate energies (1 → 1000 MeV) an adequate approximation would be

$$E_\nu \sim E_l \quad E_n \sim E_p \quad p_c^f \sim p_c^i$$

thus if we again replace our matrix element squared by $G^2$ (2.5.3) we find

$$\frac{d\sigma}{d\Omega} = \frac{G^2}{(2\pi)^2}\frac{E_\nu^2 E_n^2}{E_c^2} \tag{2.5.10}$$

$$\sigma_T = \frac{4\pi G^2}{(2\pi)^2}\frac{E_\nu^2 E_n^2}{E_c^2}$$

where $\sigma_T$ is the total cross-section. Now let us work in the laboratory system, $\sigma_T$ is Lorentz invariant and because of the massiveness of the nucleon the laboratory energies are

$$E_\nu \sim E_L \qquad E_n \sim M$$

and since (Appendix A5)

$$E_c^2 = M^2 + 2ME_L \qquad m_\nu = 0$$

we thus find

$$\sigma_T \sim G^2 \frac{E_L^2 M^2}{M^2 + 2ME_L} = \left(\frac{GM^2}{M}\right)^2 \left(\frac{E_L}{M}\right)^2 \frac{1}{1 + (2E_L/M)}$$

$$= \left(\frac{10^{-5}}{M}\right)^2 \left(\frac{E_L}{M}\right)^2 \frac{1}{1 + (2E_L/M)}. \tag{2.5.11}$$

Now the nucleon Compton wavelength is

$$\frac{1}{M} \sim 10^{-14} \text{ cm}.$$

Thus we expect

$$E_L \sim 1 \text{ MeV} \qquad \sigma_T \sim 10^{-44} \text{ cm}^2,$$

$$E_L \sim 1 \text{ GeV} \qquad \sigma_T \sim 10^{-38} \text{ cm}^2.$$

These figures agree well with experiment

Reines and Cowan (1959) $\bar{\nu}p \to ne^+$ $\qquad \sigma_T \sim 10 \times 10^{-44} \text{ cm}^2$,

Danby *et al.* (1962) $\bar{\nu}p \to n\mu^+$ $\qquad \sigma_T \sim 10^{-38} \text{ cm}^2$.

Thus the combination of phase space and a constant matrix element gives the right magnitudes for decay rates and cross-sections for a considerable number of weak interactions. However, one must not conclude from this that the matrix element is constant under all circumstances. Consider the branching ratio for the decay of pions to electrons and muons

$$R = \frac{\Gamma(\pi \to e\nu)}{\Gamma(\pi \to \mu\nu)}$$

If we make the assumption that $|M_{fi}|^2$ is constant, then $R$ is simply the ratio of phase spaces. The relevant phase-space term is given in equation (2.3.16), and since the neutrino is massless we put

$$E_\nu = |\mathbf{p}_\nu| = |\mathbf{p}_l| = p_l \qquad l = e, \mu$$

Dropping common factors we expect

$$R = \frac{\text{Phase space } \pi \to e}{\text{Phase space } \pi \to \mu}$$

$$= \frac{E_e p_e^2}{E_\mu p_\mu^2}. \tag{2.5.12}$$

Now conservation of energy and momentum gives us (with $m_\nu = 0$)

$$m_\pi = \sqrt{m_l^2 + p_l^2} + p_l$$

$$p_l = \frac{1}{2m_\pi}(m_\pi^2 - m_l^2).$$

Thus

$$R = \frac{E_e(m_\pi^2 - m_e^2)^2}{E_\mu(m_\pi^2 - m_\mu^2)^2} \sim \frac{70 \times 140^4}{100(140^2 - 100^2)^2} \sim 3. \tag{2.5.13}$$

Experimentally, the ratio is found to be $\sim 10^{-4}$ (Ashkin *et al.*, 1959). This enormous discrepancy arises from the neglect of $|M_{fi}|^2$. Before we consider matrix elements in detail, however, the problems of the next chapter must be satisfactorily settled. We shall return to the problem of $\pi$ decay in § 6.2(c).

CHAPTER 3

# RELATIVISTIC WAVE EQUATIONS AND FIELDS

IN ORDER to produce satisfactory matrix elements we must firstly produce satisfactory descriptions of particle states. These must patently be relativistic because

1. most of elementary particle physics is concerned with particles moving with velocities $\beta$ ($= v/c$) in the limit $\beta \to 1$;
2. we shall require our matrix elements to be Lorentz invariant.

This important physical property must be satisfied since we shall require the physical laws for processes to be independent of whether the reference frames are moving or stationary. With this object in mind we shall always attempt to formulate expressions as Lorentz invariants, that is they must remain invariant under Lorentz transformations. An example of such an expression would be the scalar product of two four-vectors $A_\lambda$ and $B_\lambda$; using the Lorentz transformation given in Appendix A5 it is a simple matter to demonstrate that $A_\lambda B_\lambda$ is Lorentz invariant.

We must endow our particle states with one further property—the ability to describe spin. All of these problems will be examined in the present chapter.

## 3.1. The Klein–Gordon Equation

Consider the Schrodinger equation

$$i\frac{\partial \phi}{\partial t} + \frac{1}{2m}\nabla^2 \phi = 0. \tag{3.1.1}$$

The solution

$$\phi \sim e^{ikx} = e^{i(\mathbf{k}\cdot\mathbf{x}-\omega t)}$$

then gives us

$$\omega - \frac{\mathbf{k}^2}{2m} = 0 \rightarrow \omega = \frac{\mathbf{k}^2}{2m}.$$

This is patently non-relativistic; obviously we require an equation which leads to

$$\omega^2 - \mathbf{k}^2 = m^2 \quad \text{or} \quad -k_\lambda^2 = m^2$$

if we wish to deal with relativistic systems. If we write the operator equivalent of $k_\lambda$ as

$$k_\lambda \equiv P_\lambda = -i \frac{\partial}{\partial x_\lambda}$$

then for a wave function

$$\phi \sim e^{ikx}$$

$$(P_\lambda^2 + m^2)\,\phi = (k_\lambda^2 + m^2)\,\phi = 0. \tag{3.1.2}$$

This is the Klein–Gordon equation. Since $k_\lambda^2$ is Lorentz invariant it should be possible to associate states with this equation which are independent of the Lorentz reference frame

$$P_\lambda^2\,|k_\lambda\rangle = -m^2\,|k_\lambda\rangle. \tag{3.1.3}$$

## 3.2. Particles with Spin

Now the Klein–Gordon equation is a one-component equation and so are its solutions (i.e. they consist of one term only)

$$\phi \sim e^{ikx}.$$

This is satisfactory for particles without spin, e.g. $\pi$ and $K$ mesons, since a spinless system only has one component, but if we want to go on to systems with spin, e.g. nucleons, $\varrho$-mesons, then we must add further components and have a wave function like

$$\phi_\alpha = s_\alpha e^{ikx} \tag{3.2.1}$$

where $s$ is the spin function of the system, and $\alpha$ the axis along which $s$ is projected.

Now the problems start to arise. Conventionally in non-relativistic quantum mechanics we have

$$\mathbf{S}^2 \,|s\rangle = s(s+1)\,|s\rangle \qquad S_z\,|s\rangle = s_z\,|s\rangle \tag{3.2.2}$$

where the operator $\mathbf{S}^2$ represents

$$\mathbf{S}^2 = S_x^2 + S_y^2 + S_z^2 .$$

Now this is all defined in a three-dimensional space and is apparently meaningless when we go to a relativistic system. A Lorentz transformation along the $z$-axis mixes space and time components. Similarly whilst $\mathbf{S}^2$ remains invariant under spatial rotations it is patently not invariant under Lorentz transformations.

We would therefore like to have something which is invariant under Lorentz transformations. Before reaching this goal, however, a few general remarks about the behaviour of systems under different types of transformations are needed.

## 3.3. Transformations

If we have a reference frame with coordinates $x_1, x_2, x_3, x_4$ the most general transformation we can make to it is of the form

$$x_\mu \rightarrow x'_\mu = a_\mu + a_{\mu\nu} x_\nu \tag{3.3.1}$$

where $a_\mu$ = displacement along $\mu$-axis,

$a_{\mu\nu}$ = rotation in $\mu\nu$-plane.

Now in quantum mechanics it is normal to regard the wave function

$$\psi(x) = \langle x \mid \psi \rangle \tag{3.3.2}$$

as a probability amplitude at a space-time point $x$. It is evident therefore that transformation $x \rightarrow x'$ must reflect back into the interpretation of $|\psi\rangle$.

Let us write equation (3.3.1) symbolically as

$$x \rightarrow x' = \Lambda x \tag{3.3.3}$$

then the transformation implies that the value of the wave function of $x$ is taken to $\Lambda x$ and at the same time the value of the wave function at $\Lambda^{-1}x$ moves to $x$. Thus we can write

$$\begin{array}{ll} x \xrightarrow{\Lambda} x' = \Lambda x & \psi(x) \to \psi'(x') \\ \Lambda^{-1}x \xrightarrow{\Lambda} x & \psi(\Lambda^{-1}x) \to \psi'(x) \end{array} \tag{3.3.4}$$

but if the value of the wave function remains invariant under the transformation, then

$$\psi(\Lambda^{-1}x) = \psi'(x). \tag{3.3.5}$$

Let us firstly consider displacements

$$x_\mu \to x'_\mu = x_\mu + a_\mu$$

and denote the operator which transforms $\psi$ to $\psi'$ by $U(a)$

$$|\psi'\rangle = U(a)\,|\psi\rangle. \tag{3.3.6}$$

If we assume $U(a)$ is linear in $a$ for an infinitesimal transformation ($P_\mu$ is called the generator of the displacement; its significance will emerge below)

$$U(a) = \hat{1} - ia_\mu P_\mu \qquad a_\mu \to 0 \tag{3.3.7}$$

then equations (3.3.2), (3.3.5) and (3.3.6) plus Taylor's theorem give us

$$\begin{aligned} \psi'(x) = \langle x \mid \psi' \rangle = \langle x|\,\hat{1} - ia_\mu P_\mu\,|\psi\rangle = \psi(\Lambda^{-1}x) &= \psi(x_\mu - a_\mu) \\ &= \psi(x) - a_\mu \frac{\partial \psi}{\partial x_\mu} \end{aligned}$$

therefore

$$ia_\mu \langle x|\, P_\mu\, |\psi\rangle = a_\mu \frac{\partial \psi}{\partial x_\mu}$$

$$P_\mu \equiv -i\frac{\partial}{\partial x_\mu}. \tag{3.3.8}$$

Thus we can identify the expression associated with $a_\mu$ as that for the linear four momentum operator. Physically this makes sense; if our reference frame is displaced, an observer sitting in that frame would say that a particle not associated with the frame has moved, i.e. momentum has been given to it. It is evident that a constant rate of displacement implies constant momentum, i.e. conserved momentum.

Now consider rotations and for simplicity we examine a spatial rotation in the $xy$-plane. This rotation is displayed in Fig. 3.1

$$
\begin{aligned}
x' &= x\cos\phi - y\sin\phi \overset{\phi\to 0}{\longrightarrow} x - y\,\delta\phi \\
y' &= x\sin\phi + y\cos\phi \qquad x\,\delta\phi + y \\
z' &= z \qquad\qquad\qquad\qquad z.
\end{aligned}
$$

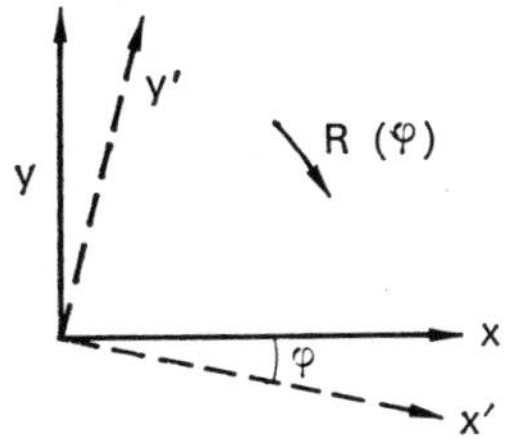


FIG. 3.1.

Thus if we represent the transformation of the state as

$$|\psi'\rangle = U(\phi)\,|\psi\rangle \tag{3.3.9}$$

and assume $U$ is linear in $\delta\phi$ for an infinitesimal transformation

$$U(\delta\phi) = \hat{1} - iJ_{xy}\,\delta\phi \qquad \delta\phi \to 0 \tag{3.3.10}$$

then as before (3.3.5)

$$
\begin{aligned}
\psi'(x) &= \psi(\Lambda^{-1}x) \\
&\downarrow \\
&\equiv x, y, z
\end{aligned}
$$

$$
\begin{aligned}
\psi'(x) &= \langle x, y, z|\,\hat{1} - iJ_{xy}\,\delta\phi\,|\psi\rangle \\
&= \psi(x, y, z) + \frac{\partial\psi}{\partial x}(y\,\delta\phi) + \frac{\partial\psi}{\partial y}(-x\,\delta\phi)
\end{aligned}
$$

$$-i\,\delta\phi\,\langle x, y, z|\,J_{xy}\,|\psi\rangle = \delta\phi\left(y\frac{\partial}{\partial x} - x\frac{\partial}{\partial y}\right)\psi$$

$$J_{xy} \equiv -i\left(x\frac{\partial}{\partial y} - y\frac{\partial}{\partial x}\right) \equiv (xP_y - yP_x). \tag{3.3.11}$$

This we recognize as the $z$-component of the orbital angular momentum operator. Thus the generator of angular rotations in the $xy$-plane is the angular momentum operator.

If the state $|\psi\rangle$ is not scalar but contains intrinsic angular momentum (i.e. spin) components $|\psi_a\rangle$, one must provide for the transformation of these components during the rotation

$$|\psi_a\rangle \to |\psi_b'\rangle = C_{ba}\,|\psi_a\rangle. \tag{3.3.12}$$

This can be done by adding a spin operator $S$ which acts on the intrinsic components only and so $J$ becomes

$$\begin{aligned} J_{xy} &= \left[-i\left(x\frac{\partial}{\partial y} - y\frac{\partial}{\partial x}\right) + S_{xy}^{ab}\right] \\ &\equiv L_z + S_z. \end{aligned} \tag{3.3.13}$$

Up to now we have ignored the fourth component—the time component—which will be necessary if we wish to go to Lorentz invariant

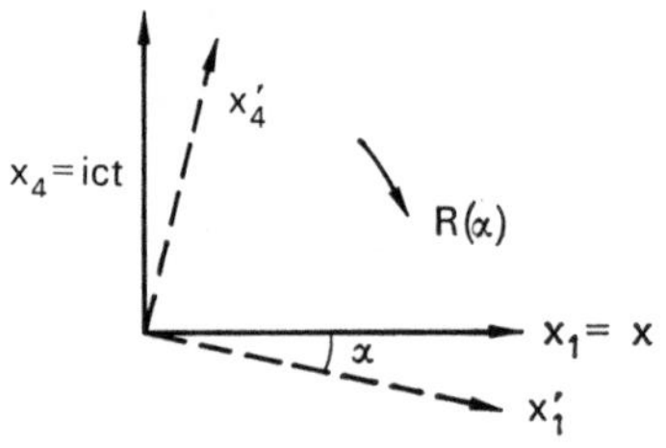


FIG. 3.2.

systems. Now if we make a rotation in say the plane of the $x$ and time axes (Fig. 3.2) we have

$$\begin{aligned} x_1' &= x_1 \cos\alpha - x_4 \sin\alpha \\ x_2' &= x_2 \qquad\qquad\qquad x_2 \equiv y \\ x_3' &= x_3 \qquad\qquad\qquad x_3 \equiv z \\ x_4' &= x_1 \sin\alpha + x_4 \cos\alpha. \end{aligned}$$

Now if we write $\tan\alpha = i\beta$, where $\beta = v/c$ = velocity, then

$$\cos\alpha = \frac{1}{\sqrt{1-\beta^2}} = \gamma \qquad \sin\alpha = i\beta\gamma \tag{3.3.14}$$

and we obtain the familiar equations for the Lorentz transformation

$$x_1' = \gamma x_1 - i\beta\gamma x_4,$$
$$x_4' = i\beta\gamma x_1 + \gamma x_4.$$

Thus rotations in the space-time planes generate Lorentz transformations. One therefore normally sets the time axis on an equal footing with the others and introduces the general operator for infinitesimal rotations [compare (3.3.10)]

$$U(\varepsilon) = \hat{1} - \frac{i}{2}\varepsilon_{\mu\nu}M_{\mu\nu} \tag{3.3.15}$$

where $\varepsilon_{\mu\nu}$ represents the rotation angle in the $\mu\nu$-plane; $M_{\mu\nu}$ incorporates both $J_{xy}$ and time terms and the factor $\frac{1}{2}$ appears because $\varepsilon_{\mu\nu} = -\varepsilon_{\nu\mu}$ and $M_{\mu\nu} = -M_{\nu\mu}$ and we sum over $\mu$, $\nu$.

Thus the full generator of displacements and rotations is

$$U(a, \varepsilon) = \hat{1} - ia_\mu P_\mu - \frac{i}{2}\varepsilon_{\mu\nu}M_{\mu\nu}. \tag{3.3.16}$$

It is called the Poincaré group of transformations. If $a$, $\varepsilon$ are finite this equation becomes

$$U(a, \varepsilon) = e^{-i(a_\mu P_\mu + \frac{1}{2}\varepsilon_{\mu\nu}M_{\mu\nu})}. \tag{3.3.17}$$

## 3.4. The Pauli–Lubanski Vector

This essentially brings us back to where we started with the problem in § 3.2. What are we going to write down for spin which makes sense for relativistic systems and transformations? We want something which is

1. Lorentz invariant;
2. commutes with $P_\lambda^2$ and $P_\lambda$ since we want to make simultaneous measurements on "spin" and momentum in order to fully define the particles properties;

3. reduces to the normal concept of spin in non-relativistic quantum mechanics in the limit $\beta \to 0$. Experience has taught us that this concept is completely successful.

The way to achieve our aim is to construct a four-vector since (1) can then be fulfilled. The simplest four-vector we can make from the spin and momentum operators is

$$X_\mu = M_{\mu\nu} P_\nu \equiv \sum_\nu M_{\mu\nu} P_\nu.$$

This gives for Lorentz transformations

$$X_\mu \to X'_\mu = a_{\mu\varrho} M_{\varrho\nu} P_\nu$$

$$X'^2_\mu = X^2_\mu$$

and so satisfies condition (1), but for $\beta \to 0$

$$P_\nu = (0, 0, 0, im)$$

and so

$$X_\mu = iM_{j4} m, 0 \qquad j = 1, 2, 3,$$

i.e. space-time generators—these do not satisfy condition (3).

The next simplest four-vector is

$$\Gamma_\sigma = \frac{1}{2i} \varepsilon_{\sigma\mu\nu\lambda} M_{\mu\nu} P_\lambda \tag{3.4.1}$$

$$\varepsilon_{\sigma\mu\nu\lambda} = \begin{cases} +1 & \text{even permutations of 1234,} \\ -1 & \text{odd permutations of 1234,} \\ 0 & \text{two indices equal.} \end{cases}$$

Since

$$M_{\mu\nu} = L_{\mu\nu} + S_{\mu\nu}$$

[compare (3.3.13)] we can drop the $L_{\mu\nu}$ component in $\Gamma_\sigma$ as each term cancels because of the antisymmetry of $\varepsilon$

$$\varepsilon_{\sigma\mu\nu\lambda} L_{\mu\nu} P_\lambda = \varepsilon_{\sigma\mu\nu\lambda}(x_\mu P_\nu - x_\nu P_\mu) P_\lambda = 0.$$

Thus

$$\Gamma_\sigma = \frac{1}{2i} \varepsilon_{\sigma\mu\nu\lambda} S_{\mu\nu} P_\lambda \tag{3.4.2}$$

and for $\beta \to 0$

$$P_\lambda = 0, 0, 0, im$$

$$\Gamma_1 = \tfrac{1}{2}(\varepsilon_{1234} S_{23} + \varepsilon_{1324} S_{32})\, m = m S_{23} \equiv m S_1 = m S_x$$

and the general term in this limit is

$$\Gamma_i = m\varepsilon_{ijk} S_{jk} = m S_i \qquad \Gamma_4 = 0 \tag{3.4.3}$$

$$\begin{aligned} \varepsilon_{ijk} &= +1 \text{ even permutations of } 123 \\ &= -1 \text{ odd permutations of } 123 \\ &= 0 \text{ two indices equal} \end{aligned}$$

and so condition (3) is satisfied. Commutation relations can also be constructed for $\Gamma_\sigma$; in the limit $\beta \to 0$ they lead to the normal commutation relations for angular momentum operators

$$[S_i, S_j] = i\varepsilon_{ijk} S_k. \tag{3.4.4}$$

Notice also that the vanishing of $L_{\mu\nu}$ makes sense; if we regard spin as the intrinsic angular momentum of a particle in its rest frame, then conventionally

$$\mathbf{J} = \mathbf{L} + \mathbf{S} = \mathbf{r} \times \mathbf{P} + \mathbf{S} \xrightarrow{\beta \to 0} \mathbf{S}.$$

Thus the normal spin operators can be regarded as non-relativistic extensions of $\Gamma_\sigma$. The eigenvalues of $\Gamma_\sigma^2$ are

$$\Gamma_\sigma^2 \, |k_\lambda = 0, im; s\rangle = m^2 \, s(s+1) \, |s\rangle \tag{3.4.5}$$

and since $\Gamma_\sigma^2$ is Lorentz invariant, this eigenvalue remains the same in any reference frame. In the rest frame of the particle

$$\mathbf{\Gamma} = m\mathbf{S} \qquad \Gamma_4 = 0.$$

Now consider the situation when $m \to 0$ we then have

$$\Gamma_\sigma^2 \to 0 \qquad P_\sigma^2 \to 0 \quad \text{for} \quad m \to 0. \tag{3.4.6}$$

But we also have

$$\Gamma_\sigma P_\sigma = 0 \tag{3.4.7}$$

since for a particle of finite mass in its rest system

$$\Gamma_\sigma = (m\mathbf{S}, 0) \qquad P_\sigma = (0, im)$$

and so

$$\Gamma_\sigma P_\sigma = 0.$$

This is a Lorentz invariant expression and must hold in all reference frames, even when $\beta \to 1$ which is equivalent to saying $m \to 0$.

Thus we have the three equations

$$\Gamma_\sigma^2 = 0 \qquad P_\sigma^2 = 0 \qquad \Gamma_\sigma P_\sigma = 0$$

and can deduce that

$$\Gamma_\sigma = \lambda P_\sigma \tag{3.4.8}$$

$$\mathbf{\Gamma} = \lambda \mathbf{P} \qquad \Gamma_4 = \lambda P_4$$

where $\lambda$ is a term whose structure we must examine. Consider a common Lorentz transformation for the four-vectors $p$ and $\Gamma$

$$p'_\mu = L_{\mu\nu} p_\nu \qquad \Gamma'_\mu = L_{\mu\nu}\Gamma_\nu$$

where $L_{\mu\nu}$ is the Lorentz transformation matrix (Appendix A5). If we start with a system of finite mass at rest we then have for a transformation along the $x$-axis

$$\begin{pmatrix} p'_1 \\ p'_2 \\ p'_3 \\ p'_4 \end{pmatrix} = \begin{pmatrix} \gamma & 0 & 0 & -i\beta\gamma \\ 0 & 1 & 0 & 0 \\ 0 & 0 & 1 & 0 \\ i\beta\gamma & 0 & 0 & \gamma \end{pmatrix} \begin{pmatrix} 0 \\ 0 \\ 0 \\ im \end{pmatrix} = \begin{pmatrix} \beta\gamma m \\ 0 \\ 0 \\ i\gamma m \end{pmatrix}$$

$$\begin{pmatrix} \Gamma'_1 \\ \Gamma'_2 \\ \Gamma'_3 \\ \Gamma'_4 \end{pmatrix} = \begin{pmatrix} \gamma & 0 & 0 & -i\beta\gamma \\ 0 & 1 & 0 & 0 \\ 0 & 0 & 1 & 0 \\ i\beta\gamma & 0 & 0 & \gamma \end{pmatrix} \begin{pmatrix} mS\cos\vartheta \\ mS\sin\vartheta \\ 0 \\ 0 \end{pmatrix} = \begin{pmatrix} \gamma mS\cos\vartheta \\ mS\sin\vartheta \\ 0 \\ i\beta\gamma\, mS\cos\vartheta \end{pmatrix}$$

where we have simplified calculation by assuming $\mathbf{\Gamma}$ lies in the $xy$-plane. Now in the limit $m \to 0$

$$m \to 0 \quad \beta \to 1 \quad \gamma \to \infty$$

$$m\beta\gamma \to m\gamma = E = |\mathbf{p}|$$

$$\frac{\Gamma'_2}{\Gamma'_1} = \frac{1}{\gamma}\tan\vartheta \to 0$$

that is the components of $\Gamma$ perpendicular to $\mathbf{p}$ vanish; this is the result expected from (3.4.8). In addition in the limit $m \to 0$

$$\frac{\Gamma_1'}{p_1'} = \frac{\gamma m S \cos\vartheta}{\beta\gamma m} \xrightarrow{m\to 0} \frac{\mathbf{S}\cdot\mathbf{p}}{|\mathbf{p}|}$$

$$\frac{\Gamma_4'}{p_4'} = \frac{i\beta\gamma m S \cos\vartheta}{i\gamma m} \xrightarrow{m\to 0} \frac{\mathbf{S}\cdot\mathbf{p}}{|\mathbf{p}|}$$

thus from (3.4.8)

$$\lambda = \frac{\mathbf{S}\cdot\mathbf{p}}{|\mathbf{p}|}. \tag{3.4.9}$$

It is apparent that $\lambda$ is the same in all Lorentz frames for $m \to 0$. It is called the helicity operator and for a massless particle the eigenvalue of $|\lambda|$ can be regarded as its spin value. For particles with definite parity $\lambda$ can take on two values since for a spatial reflection [compare § 6.1 (a)]

$$\mathbf{S} \xrightarrow{\mathbf{x}\to -\mathbf{x}} \mathbf{S} \qquad \mathbf{p} \xrightarrow{\mathbf{x}\to -\mathbf{x}} -\mathbf{p}$$

$$\mathbf{S}\cdot\mathbf{p} \to -\mathbf{S}\cdot\mathbf{p}$$

and therefore

$$\lambda = \pm\, |\lambda|. \tag{3.4.10}$$

For particles with no specific parity only one value of $|\lambda|$ exists and so it always has a definite helicity. This situation arises for neutrinos and we shall examine it in Chapter 6.

Note that for particles of finite mass one can define a helicity operator

$$\begin{aligned}
\lambda(p) &= \frac{i}{m^2}\left(\Gamma_4\, |\mathbf{p}| - \mathbf{\Gamma}\cdot\mathbf{p}\frac{p_4}{|\mathbf{p}|}\right) \\
&= \frac{i}{m^2}\left(i\beta\gamma m S\cos\vartheta\, \beta\gamma m - \gamma m S \cos\vartheta\, \gamma\beta m \frac{i\gamma m}{\beta\gamma m}\right) \\
&= -\gamma^2(\beta^2 - 1)\, S\cos\vartheta \\
&= \frac{\mathbf{S}\cdot\mathbf{p}}{|\mathbf{p}|}
\end{aligned} \tag{3.4.11}$$

since $\lambda(p)$ is independent of $\beta$ and $\gamma$ it is therefore independent of Lorentz transformations in direction of $\mathbf{p}$. If we choose the $\mathbf{p}$-axis as axis of quantisation then the eigenvalues of $\lambda(p)$ are $-s,\ -s+1,\ \ldots,\ +s$ on states $|\,p, s, \lambda\rangle$.

In limit $m \to 0$ we then find

$$\lambda(p) \to \frac{i}{m^2}(-i\Gamma_4 p_4 - i\mathbf{\Gamma}\cdot\mathbf{p}) = \frac{1}{m^2}\Gamma p = \frac{\lambda}{m^2}p^2 = -\lambda.$$

## 3.5. Wave Equations for Particles of Non-zero Spin

### 3.5(a). The Dirac equation

Historically the Dirac equation arose from certain difficulties in the interpretation of the Klein–Gordon equation. Since they are not relevant to the present discussion we shall not elaborate on them further beyond saying that they occurred because the Klein–Gordon equation involved second-order differentials (the interested reader can pursue the matter further on p. 55 of *P.E.P.*). Dirac tried to avoid this problem by constructing a covariant expression using the operator $\partial/\partial x_\lambda$. In order to make a Lorentz invariant equation this operator must then be combined with another four vector to yield a scalar product; this we shall write as

$$\gamma_\lambda \frac{\partial}{\partial x_\lambda} = \sum_{\lambda=1}^{4} \gamma_\lambda \frac{\partial}{\partial x_\lambda}.$$

This is a scalar operator under Lorentz transformations but the possibility exists that the desired equation contains another scalar operator not involving $\partial/\partial x_\lambda$; thus its most general form is

$$\left(\gamma_\lambda \frac{\partial}{\partial x_\lambda} + c\right)\psi = 0$$

where $c$ is a scalar operator. We wish this equation to satisfy the Einstein relation (2.1.1)

$$p_\lambda^2 = -m^2 \equiv \left(\frac{\partial^2}{\partial x_\lambda^2} - m^2\right)\psi = 0 \qquad (3.5.1)$$

which can be done by operating in the left by

$$\left(\gamma_\mu \frac{\partial}{\partial x_\mu} - c\right)$$

then

$$\left(\gamma_\mu \frac{\partial}{\partial x_\mu} - c\right)\left(\gamma_\lambda \frac{\partial}{\partial x_\lambda} + c\right)\psi = 0,$$

$$\left(\gamma_\mu \gamma_\lambda \frac{\partial}{\partial x_\mu} \frac{\partial}{\partial x_\lambda} - c^2\right)\psi = 0.$$

Equation (3.5.1) is then satisfied if

$$\gamma_\mu \gamma_\lambda + \gamma_\lambda \gamma_\mu = 2\delta_{\mu\lambda} \tag{3.5.2}$$

$$c = m\hat{1}.$$

Thus the Dirac equation is

$$\left(\gamma_\lambda \frac{\partial}{\partial x_\lambda} + m\right)\psi = 0 \tag{3.5.3}$$

where we have dropped the unit operator since we shall have no need to display it explicitly in future.

The equation (3.5.2) implies that

1. $\gamma$ is a matrix,
2. no unique solution is possible for $\gamma$ since this is the only information we have on its properties,
3. the multiplication of the $\gamma$-matrix with itself yields sixteen independent matrices. They are as follows:

| | | number of matrices |
|---|---|---|
| $\hat{1}$ | unit matrix | 1 |
| $\gamma_\lambda$ | | 4 |
| $\gamma_\lambda \gamma_\mu$ | $\lambda < \mu$ | 6 |
| $\gamma_\lambda \gamma_\mu \gamma_\nu$ | $\lambda < \mu < \nu$ | 4 |
| $\gamma_1 \gamma_2 \gamma_3 \gamma_4 = \gamma_5$ | | 1. |

It is easy to verify that this is the maximum number of independent matrices; if we tried, say, the combination $\gamma_1\gamma_2\gamma_3\gamma_4\gamma_2$, then by (3.5.2)

$$\gamma_1\gamma_2\gamma_3\gamma_4\gamma_2 = -\gamma_1\gamma_2\gamma_3\gamma_2\gamma_4 = \gamma_1\gamma_2\gamma_3\gamma_4 = \gamma_1\gamma_3\gamma_4.$$

Although no unique solution is available for the $\gamma$-matrix, a set of matrices with sixteen independent values must be at least $4 \times 4$ in size. A common representation used for the $\gamma$-matrices is the Dirac–Pauli set

$$\gamma_k = \begin{pmatrix} 0 & -i\sigma_k \\ i\sigma_k & 0 \end{pmatrix} \quad \gamma_4 = \begin{pmatrix} 1 & 0 \\ 0 & -1 \end{pmatrix} \quad \gamma_5 = \begin{pmatrix} 0 & -1 \\ -1 & 0 \end{pmatrix} \tag{3.5.4}$$

where $k = 1, 2, 3$ and each symbol is a $2 \times 2$ matrix

$$\sigma_1 = \begin{pmatrix} 0 & 1 \\ 1 & 0 \end{pmatrix} \quad \sigma_2 = \begin{pmatrix} 0 & -i \\ i & 0 \end{pmatrix} \quad \sigma_3 = \begin{pmatrix} 1 & 0 \\ 0 & -1 \end{pmatrix} \tag{3.5.5}$$

$$1 = \begin{pmatrix} 1 & 0 \\ 0 & 1 \end{pmatrix} \quad 0 = \begin{pmatrix} 0 & 0 \\ 0 & 0 \end{pmatrix}.$$

With this choice of representation the $\gamma$-matrices are hermitian

$$\gamma_\lambda = \gamma_\lambda^\dagger. \tag{3.5.6}$$

It is often convenient to express the sixteen independent combinations in terms of $\gamma_5$; we then have

$$\Gamma = \begin{matrix} 1 & & \text{S} \\ \gamma_\lambda & & \text{V} \\ i\gamma_\lambda\gamma_\mu & \lambda < \mu & \text{T} \\ i\gamma_\lambda\gamma_5 & & \text{A} \\ \gamma_5 & & \text{P} \end{matrix} \tag{3.5.7}$$

The presence of the $i$ ensures that $\Gamma$ is hermitian and the letters S, V, T, A, P imply that the $\Gamma$s behave like scalars, vectors, tensors, axial vectors and pseudoscalars under Lorentz transformations (see p. 85 of *P.E.P.* for a proof). Other useful relations involving $\gamma$-matrices are

$$\gamma_5\gamma_\lambda + \gamma_\lambda\gamma_5 = 2\delta_{\lambda 5} \tag{3.5.8}$$

$$\sigma_{\mu\nu} = \underset{\mu \neq \nu}{-i\gamma_\mu\gamma_\nu} = \frac{1}{2i}(\gamma_\mu\gamma_\nu - \gamma_\nu\gamma_\mu). \tag{3.5.9}$$

### 3.5(b). Adjoint form of the Dirac equation

Consider the Dirac equation (3.5.3)

$$\left(\gamma_\lambda \frac{\partial}{\partial x_\lambda} + m\right)\psi = 0$$

since $\gamma$ is a matrix so also is $\psi$, therefore if we take the adjoint form and recall that for two matrices $(AB)^\dagger = B^\dagger A^\dagger$, then

$$\left(\frac{\partial \psi}{\partial x_\lambda}\right)^\dagger \gamma_\lambda^\dagger + m\psi^\dagger = \left(\frac{\partial \psi}{\partial x_\lambda}\right)^\dagger \gamma_\lambda + m\psi^\dagger = 0$$

where we have used the Hermitian property of $\gamma_\lambda$ (3.5.6). If we multiply this equation from the right-hand side by $\gamma_4$ and introduce the definition

$$\bar{\psi} = \psi^\dagger \gamma_4 \tag{3.5.10}$$

we obtain the adjoint form of the Dirac equation

$$\frac{\partial \bar{\psi}}{\partial x_\lambda}\gamma_\lambda - m\bar{\psi} = 0. \tag{3.5.11}$$

### 3.5(c). Currents for Dirac particles

Let us first recall the continuity equation for fluids

$$\frac{\partial \varrho}{\partial t} + \nabla \cdot \mathbf{j} = 0 \tag{3.5.12}$$

where $\varrho$ and $j$ represent flux and current density respectively. This equation is simply a statement of the conservation of material ($j_n$ represents the flux normal to the surface element $d\sigma$)

$$\frac{\partial}{\partial t}\int\limits_V \varrho \, d\mathbf{x} = -\int\limits_A j_n \, d\sigma = -\int\limits_V \nabla \cdot \mathbf{j} \, d\mathbf{x}.$$

In four-vector notation equation (3.5.12) reduces to

$$\frac{\partial j_\lambda}{\partial x_\lambda} = 0 \tag{3.5.13}$$

and this equation is satisfied for Dirac particles if we define the four-vector current density as

$$j_\lambda = i\bar{\psi}\gamma_\lambda\psi \tag{3.5.14}$$

then by (3.5.3) and (3.5.11)

$$i\frac{\partial}{\partial x_\lambda}(\bar{\psi}\gamma_\lambda\psi) = i\left(\frac{\partial\bar{\psi}}{\partial x_\lambda}\gamma_\lambda\psi + \bar{\psi}\gamma_\lambda\frac{\partial\psi}{\partial x_\lambda}\right)$$
$$= i(m\bar{\psi}\psi - m\bar{\psi}\psi) = 0;$$

furthermore, the fourth component yields the conventional quantum mechanical definition of probability density $\varrho$

$$\varrho = \frac{1}{i}j_4 = \bar{\psi}\gamma_4\psi = \psi^\dagger\gamma_4\gamma_4\psi = |\psi|^2$$

which is a positive definite quantity.

### 3.5(d). Spin operators for Dirac particles

Consider a Lorentz transformation of the type

$$x_\mu \to x'_\mu = a_{\mu\nu}x_\nu \tag{3.5.15}$$

then since scalar products remain invariant under Lorentz transformations

$$x'^2_\mu = a_{\mu\nu}a_{\mu\varrho}x_\nu x_\varrho$$
$$= x^2_\nu = \delta_{\nu\varrho}x_\nu x_\varrho .$$

Thus

$$a_{\mu\nu}a_{\mu\varrho} = \delta_{\nu\varrho} \tag{3.5.16}$$

and the reverse transformation can also be obtained

$$x'_\mu = a_{\mu\nu}x_\nu$$

$$a_{\mu\varrho}x'_\mu = a_{\mu\varrho}a_{\mu\nu}x_\nu = \delta_{\varrho\nu}x_\nu = x_\varrho$$

$$x_\varrho = a_{\mu\varrho}x'_\mu . \tag{3.5.17}$$

Now consider the Dirac equation

$$\left(\gamma_\mu \frac{\partial}{\partial x_\mu} + m\right)\psi(x) = 0 \tag{3.5.18}$$

and make a Lorentz transformation $x_\mu \to x'_\mu$ giving $\psi'(x') = U\psi(x)$; we then have

$$\left(\gamma_\mu \frac{\partial}{\partial x'_\mu} + m\right)\psi'(x') = \left(\gamma_\mu \frac{\partial}{\partial x'_\mu} + m\right) U\psi(x) = 0$$

but by (3.5.17)

$$\frac{\partial}{\partial x'_\mu} = \frac{\partial x_\nu}{\partial x'_\mu}\frac{\partial}{\partial x_\nu} = a_{\mu\nu}\frac{\partial}{\partial x_\nu} \tag{3.5.19}$$

and if we multiply the left-hand side of our equation by $U^{-1}$ we obtain

$$\left(U^{-1}\gamma_\mu U a_{\mu\nu}\frac{\partial}{\partial x_\nu} + m\right)\psi(x) = 0.$$

We have now completed the reverse transformation and comparison with (3.5.18) then gives us

$$U^{-1}\gamma_\mu U a_{\mu\nu} = \gamma_\nu. \tag{3.5.20}$$

If we multiply both sides by $a_{\lambda\nu}$ and use equation (3.5.16), we obtain

$$U^{-1}\gamma_\lambda U = a_{\lambda\nu}\gamma_\nu.$$

Now consider $a_{\mu\nu}$ to be explicitly of the form

$$x'_\mu = a_{\mu\nu}x_\nu = (\delta_{\mu\nu} - \varepsilon_{\mu\nu})\, x_\nu \qquad \varepsilon \to 0 \tag{3.5.21}$$

where $\varepsilon_{\mu\nu}$ is a small rotation; the corresponding transformation operator for $\psi$ is

$$U = \hat{1} - \frac{i}{2}\varepsilon_{\mu\nu}S_{\mu\nu} \tag{3.5.22}$$

where the spatial components of $S_{\mu\nu}$ represent the spin operator [compare (3.3.15) and our discussion in § 3.3]. Using the above operator our equation

$$U^{-1}\gamma_\lambda U = a_{\lambda\nu}\gamma_\nu$$

becomes with neglect of terms of order $\varepsilon^2$

$$\left(\hat{1} + \frac{i}{2}\varepsilon_{\mu\nu}S_{\mu\nu}\right)\gamma_\lambda\left(\hat{1} - \frac{i}{2}\varepsilon_{\mu\nu}S_{\mu\nu}\right) = (\delta_{\lambda\nu} - \varepsilon_{\lambda\nu})\,\gamma_\nu$$

$$-\frac{i}{2}\varepsilon_{\mu\nu}[S_{\mu\nu}, \gamma_\lambda] = \varepsilon_{\lambda\nu}\gamma_\nu$$

$$= \frac{1}{2}(\varepsilon_{\lambda\nu} - \varepsilon_{\nu\lambda})\,\gamma_\nu$$

$$= \frac{1}{2}\varepsilon_{\mu\nu}(\delta_{\mu\lambda}\gamma_\nu - \delta_{\nu\lambda}\gamma_\mu)$$

$$-i[S_{\mu\nu}, \gamma_\lambda] = \delta_{\mu\lambda}\gamma_\nu - \delta_{\nu\lambda}\gamma_\mu.$$

Direct substitution then shows that the solution of this equation is

$$S_{\mu\nu} = -\frac{i}{2}\gamma_\mu\gamma_\nu \qquad \mu \neq \nu \tag{3.5.23}$$

and so the space components are

$$S_1 \equiv S_{23} = -\frac{i}{2}\gamma_2\gamma_3 \quad S_2 \equiv S_{31} = -\frac{i}{2}\gamma_3\gamma_1 \quad S_3 \equiv S_{12} = -\frac{i}{2}\gamma_1\gamma_2 \tag{3.5.24}$$

then by (3.2.2) and (3.5.2)

$$\mathbf{S}^2 = S_1^2 + S_2^2 + S_3^2 = \left(\frac{1}{4} + \frac{1}{4} + \frac{1}{4}\right)\hat{1} = \frac{1}{2}\left(\frac{1}{2} + 1\right)\hat{1}. \tag{3.5.25}$$

Thus the Dirac equation is suitable for spin $\frac{1}{2}$ particles.‡ One also finds that

$$(\mathbf{S}\cdot\mathbf{p})^2 = (S_i p_i)\,(S_j p_j) = \frac{1}{4}\delta_{ij}p_i p_j = \frac{1}{4}\mathbf{p}^2 \tag{3.5.26}$$

and so the operator $\mathbf{S}\cdot\mathbf{p}$ must have eigenvalues $\pm\frac{1}{2}\,|\mathbf{p}|$. The eigenfunctions for $\mathbf{S}\cdot\mathbf{p}/|\mathbf{p}|$ are the helicity states of the Dirac particle [compare (3.4.9) and (3.4.10)].

‡ Since the eigenvalues of $\mathbf{S}^2$ are $s(s + 1)$.

### 3.5(e). Solutions of the Dirac equation

Since relativistic wave equations must satisfy the condition [compare (3.1.2)]

$$(P_\lambda^2 + m^2)\Psi = 0 \qquad \Psi \equiv \phi, \psi$$

they inevitably carry with them two solutions for the energy states

$$E = \pm\sqrt{p^2 + m^2}. \tag{3.5.27}$$

The interpretation of negative energy states caused considerable worry at the time that the Klein–Gordon and Dirac equations were introduced. Dirac explained the negative energy states as energy levels which were normally filled, but when a particle was removed from this system, the hole created had the same properties as a particle with opposite charge and positive energy.

The argument is probably more easily seen in modern language. Consider a particle (say the electron) and antiparticle (positron) with opposite properties

| | | | | |
|---|---|---|---|---|
| particle | $-e$ | $-\mathbf{p}$ | $-E$ | $-\mathbf{s}$ |
| antiparticle | $+e$ | $+\mathbf{p}$ | $+E$ | $+\mathbf{s}$ |

Now apply this situation to muon decay and a leptonic interaction

$$\mu^+ \to e^+\nu\bar{\nu}$$

$$\hookrightarrow e^-\mu^+ \to \nu\bar{\nu}$$

If the electron has all the opposite properties to the positron then the kinematic conditions of the $\nu\bar{\nu}$ pair are unaltered. Thus it is immaterial whether we consider the creation of a particle or the annihilation of an antiparticle with opposite properties.

Therefore in considering the Dirac equation we start with two solutions

$$\begin{array}{ll} \psi = u\, e^{ipx} & \psi = v\, e^{-ipx} \\ +\mathbf{p},\, +E & -\mathbf{p},\, -E \end{array} \tag{3.5.28}$$

where $u$ and $v$ are called the Dirac spinors. Substitution in the Dirac equation (3.5.3) gives

$$\begin{aligned} (i\gamma p + m)\, u &= 0 \\ (i\gamma p - m)\, v &= 0. \end{aligned} \tag{3.5.29}$$

Consider first the limit $\mathbf{p} \to 0$, then $p_4 \to im$, $u \to u_R$ and our equations become

$$\gamma_4 u_R = \hat{1}\, u_R \qquad \gamma_4 v_R = -\hat{1}\, v_R.$$

Since

$$\gamma_4 = \begin{pmatrix} 1 & 0 & 0 & 0 \\ 0 & 1 & 0 & 0 \\ 0 & 0 & -1 & 0 \\ 0 & 0 & 0 & -1 \end{pmatrix} \qquad \hat{1} = \begin{pmatrix} 1 & 0 & 0 & 0 \\ 0 & 1 & 0 & 0 \\ 0 & 0 & 1 & 0 \\ 0 & 0 & 0 & 1 \end{pmatrix}$$

we obtain four solutions

$$u_{1R} = \begin{pmatrix} 1 \\ 0 \\ 0 \\ 0 \end{pmatrix} \quad u_{2R} = \begin{pmatrix} 0 \\ 1 \\ 0 \\ 0 \end{pmatrix} \quad v_{1R} = \begin{pmatrix} 0 \\ 0 \\ 1 \\ 0 \end{pmatrix} \quad v_{2R} = \begin{pmatrix} 0 \\ 0 \\ 0 \\ 1 \end{pmatrix} \tag{3.5.30}$$

These can be interpreted as spin-up, spin-down solutions, in fact operating with $S_z = S_3$ (3.5.24) we find

$$\begin{aligned} S_z u_{1R} &= +\tfrac{1}{2} u_{1R} \qquad & S_z v_{1R} &= +\tfrac{1}{2} v_{1R} \\ S_z u_{2R} &= -\tfrac{1}{2} u_{2R} \qquad & S_z v_{2R} &= -\tfrac{1}{2} v_{2R}. \end{aligned} \tag{3.5.31}$$

Now consider solutions when $\mathbf{p} \nrightarrow 0$. The space part $e^{ipx}$ is patently Lorentz invariant. The appropriate form for $u$ can be obtained from $u_R$ by a Lorentz transformation (boost) by a rotation in the space-time planes [compare (3.3.9) and (3.3.17)]

$$u(\mathbf{p}) = U(\mathbf{p})\, u_R = e^{-i\varepsilon_{\mu\nu} S_{\mu\nu}/2}\, u_R \tag{3.5.32}$$

with $\nu = 4$. At this point one has to go a little carefully. The factor $\frac{1}{2}$ was introduced originally for convenience of summation over both indices. It is now convenient to retain it for reasons which will become apparent below. Consider a boost in the $x$ direction, i.e. a rotation in the 14 plane; we then have with the aid of (3.5.23)

$$-\frac{i}{2}(\varepsilon_{14} S_{14} + \varepsilon_{41} S_{41}) = -\frac{i}{2}(2\varepsilon_{14} S_{14}) = \frac{i}{2}\varepsilon_{14} i\gamma_1\gamma_4.$$

Thus if $\varepsilon_{14}$ is finite

$$
\begin{aligned}
U &= e^{i/2\varepsilon_{14}i\gamma_1\gamma_4} \\
&= \hat{1} + i(i\gamma_1\gamma_4)\frac{\varepsilon_{14}}{2} - \frac{1}{2!}\left(\frac{\varepsilon_{14}}{2}\right)^2 - \frac{i}{3!}(i\gamma_1\gamma_4)\left(\frac{\varepsilon_{14}}{2}\right)^3 + \cdots \\
&= \cos\frac{\varepsilon_{14}}{2}\left[\hat{1} + i(i\gamma_1\gamma_4)\tan\frac{\varepsilon_{14}}{2}\right] \\
&= \sqrt{\frac{1+\cos\varepsilon_{14}}{2}}\left[\hat{1} + i(i\gamma_1\gamma_4)\frac{\sin\varepsilon_{14}}{1+\cos\varepsilon_{14}}\right].
\end{aligned}
\tag{3.5.33}
$$

Now from equations (3.5.4) and (3.3.14)

$$i\gamma_1\gamma_4 = i\begin{pmatrix} 0 & -i\sigma_1 \\ i\sigma_1 & 0 \end{pmatrix}\begin{pmatrix} 1 & 0 \\ 0 & -1 \end{pmatrix} = \begin{pmatrix} 0 & -\sigma_1 \\ -\sigma_1 & 0 \end{pmatrix}$$

$$\cos\varepsilon_{14} = \gamma = \frac{E}{m} \qquad \sin\varepsilon_{14} = i\beta_1\gamma = \frac{ip_1}{m}$$

hence

$$U = \sqrt{\frac{m+E}{2m}}\left[\begin{pmatrix} 1 & 0 \\ 0 & 1 \end{pmatrix} + \frac{1}{m+E}\begin{pmatrix} 0 & \sigma_1 p_1 \\ \sigma_1 p_1 & 0 \end{pmatrix}\right].$$

It is apparent that if we extend the above argument to all three axes we shall obtain

$$U = \sqrt{\frac{m+E}{2m}}\left[\begin{pmatrix} 1 & 0 \\ 0 & 1 \end{pmatrix} + \frac{1}{m+E}\begin{pmatrix} 0 & \boldsymbol{\sigma}\cdot\mathbf{p} \\ \boldsymbol{\sigma}\cdot\mathbf{p} & 0 \end{pmatrix}\right]. \tag{3.5.34}$$

The relevant spinors (3.5.32) are then as follows:

(3.5.35)

| $u_1/N$ | $u_2/N$ | $v_1/N$ | $v_2/N$ |
|---|---|---|---|
| $1$ | $0$ | $\frac{p_z}{E+m}$ | $\frac{p_x - ip_y}{E+m}$ |
| $0$ | $1$ | $\frac{p_x + ip_y}{E+m}$ | $\frac{-p_z}{E+m}$ |
| $\frac{p_z}{E+m}$ | $\frac{p_x - ip_y}{E+m}$ | $1$ | $0$ |
| $\frac{p_x + ip_y}{E+m}$ | $\frac{-p_z}{E+m}$ | $0$ | $1$ |

where

$$N = \sqrt{\frac{E+m}{2m}}.$$

It is apparent that in the limit $\mathbf{p} \to 0$ we recover our original spinors (3.5.30). The spinors given above are not eigenstates of $S_z$ except when $p_x = p_y = 0$; only for $\mathbf{p} \to 0$ do they reduce to eigenstates of $S_z$ in general [compare (3.5.31)]. This is the result to be expected since we have seen in § 3.4 that the conventional concept of spin only has a well-defined meaning in the rest frame of a particle.

### 3.5(f). Dirac equation for m = 0

In the situation $m \to 0$ the spinor values in (3.5.35) apparently $\to \infty$. In fact this is not so as one can produce normalisation requirements which avoid this problem. In any case as we have shown previously the ordinary concept of spin has not got much meaning as velocity $v \to c$. Consider our spinor equations (3.5.29)

$$(i\gamma p + m)\, u = 0$$

$$(i\gamma p - m)\, v = 0.$$

For $m = 0$ we combine these equations into one for a single spinor $u_\nu$ (where the subscript $\nu$ indicates neutrino, the only known Dirac particle with zero mass)

$$i\gamma p u_\nu = 0$$

$$i\boldsymbol{\gamma} \cdot \mathbf{p}\, u_\nu = -i\gamma_4 p_4 u_\nu. \qquad (3.5.36)$$

Now recall from (3.5.24) that

$$S_3 = -\frac{i}{2}\gamma_1\gamma_2$$

then using the properties of the $\gamma$-matrices [§ 3.5(a)] it is not difficult to show that

$$S_3 = -\frac{i}{2}\gamma_1\gamma_2 = \frac{i}{2}\gamma_4\gamma_5\gamma_3$$

hence ($k = 1, 2, 3$)

$$S_k = \frac{i}{2}\gamma_4\gamma_5\gamma_k \qquad \mathbf{S} = \frac{i}{2}\gamma_4\gamma_5\boldsymbol{\gamma}. \tag{3.5.37}$$

Thus equation (3.5.36) can be rewritten as

$$i\gamma_4\gamma_5\boldsymbol{\gamma}\cdot\mathbf{p}\,u_\nu = 2\mathbf{S}\cdot\mathbf{p}\,u_\nu = -i\gamma_4\gamma_5\gamma_4p_4u_\nu = -\gamma_5 Eu_\nu$$

$$\mathbf{S}\cdot\mathbf{p}\,u_\nu = -\tfrac{1}{2}\gamma_5 Eu_\nu$$

and multiplying both sides by $\gamma_5$ yields

$$\mathbf{S}\cdot\mathbf{p}\,\gamma_5 u_\nu = -\tfrac{1}{2}\,Eu_\nu.$$

Since $E = |\mathbf{p}|$ for $m = 0$ linear combinations of the above equations yield

$$\frac{\mathbf{S}\cdot\mathbf{p}}{|\mathbf{p}|}(1+\gamma_5)\,u_\nu = -\tfrac{1}{2}(1+\gamma_5)\,u_\nu \tag{3.5.38}$$

$$\frac{\mathbf{S}\cdot\mathbf{p}}{|\mathbf{p}|}(1-\gamma_5)\,u_\nu = \tfrac{1}{2}(1-\gamma_5)\,u_\nu.$$

Thus $(1 \pm \gamma_5)\,u_\nu$ are helicity eigenstates of $\mathbf{S}\cdot\mathbf{p}/|\mathbf{p}|$ [compare (3.4.9)], and once again we see that the spin of a Dirac particle is $\frac{1}{2}\,\hbar$. Experiments, which we shall discuss in Chapter 6, have shown that only the negative helicity states of the neutrino occur in nature. The choice $E = -|\mathbf{p}|$ for equations (3.5.38) would then imply that we should expect only positive helicity states for the antineutrino, which can be regarded as the negative energy state of the neutrino [compare § 3.5(e)]; experiment confirms this expectation.

### 3.5(g). Projection operators for Dirac spinors

Consider the spinor equations (3.5.29)

$$(i\gamma p + m)\,u = 0$$

$$(i\gamma p - m)\,v = 0$$

and introduce operators

$$\Lambda^{\mp} = \frac{\pm i\gamma p + m}{2m} \tag{3.5.39}$$

then

$$\Lambda^+ u = \frac{(-i\gamma p + m)}{2m} u = \frac{2mu}{2m} = u \tag{3.5.40}$$

similarly

$$\Lambda^- u = 0 \qquad \Lambda^+ v = 0 \qquad \Lambda^- v = v. \tag{3.5.41}$$

The $\Lambda$s are therefore projection operators and satisfy the following properties

$$\Lambda^+ + \Lambda^- = \hat{1} \qquad (\Lambda^\pm)^2 = \Lambda^\pm \qquad \Lambda^\pm \Lambda^\mp = 0. \tag{3.5.42}$$

In addition by considering explicit forms of the Dirac spinors one finds

$$\Lambda^+ = \sum_{r=1}^{2} u_r \bar{u}_r \qquad \Lambda^- = -\sum_{r=1}^{2} v_r \bar{v}_r \tag{3.5.43}$$

where the subscript $r$ refers to the states listed in (3.5.35).

### 3.5(h). Sums over spin states

In considering the matrix elements for transition probabilities when Dirac particles are involved, one finds terms of the type

$$\bar{u}_{r'}\, 0 u_r \equiv \begin{cases} M_{fi} \\ T_{fi} \end{cases}$$

where 0 is a combination of $\gamma$-matrices; hence using $\bar{u}_{r'} = u^\dagger_{r'}\gamma_4$ we can write the sum over spin states for transition probabilities as

$$\begin{aligned}
\sum_{i,f} |T_{fi}|^2 &= \sum_{i,f} \langle f|\, T\, |i\rangle \langle f|\, T\, |i\rangle^* \\
&= \sum_{i,f} \langle f|\, T\, |i\rangle \langle i|\, T^\dagger\, |f\rangle \\
&\equiv \sum_{r,r'} (\bar{u}_{r'}\, 0 u_r)(u_r^\dagger\, 0^\dagger \gamma_4^\dagger u_{r'}) \\
\gamma_4^2 = \hat{1}, \quad &= \sum_{r,r'} (\bar{u}_{r'}\, 0 u_r \bar{u}_r \gamma_4 0^\dagger \gamma_4 u_{r'}) \\
&= \sum_{r'} (\bar{u}_{r'}\, 0 \Lambda^+ \gamma_4 0^\dagger \gamma_4 u_{r'}) \\
\tilde{0} = \gamma_4 0^\dagger \gamma_4, \quad &= \sum_{r'} (\bar{u}_{r'} 0 \Lambda^+ \tilde{0} \Lambda^{+\prime} u_{r'} - \bar{v}_{r'} 0 \Lambda^+ \tilde{0} \Lambda^{+\prime} v_{r'})
\end{aligned}$$

since [§ 3.5(g)]

$$\Lambda^{+\prime} u_{r'} = u_{r'} \qquad \Lambda^{+\prime} v_{r'} = 0.$$

Using explicit forms of Dirac spinors one finds for any $4 \times 4$ matrix $Q$

$$\sum_{r'} (\bar{u}_{r'} Q u_{r'} - \bar{v}_{r'} Q v_{r'}) = \operatorname{tr} Q$$

where $\operatorname{tr} Q$ is the trace of the matrix $Q$, hence

$$\begin{aligned} \sum_{r,r'} |\bar{u}_{r'}\, 0 u_r|^2 &= \operatorname{tr}\,(0 \Lambda^+ \gamma_4 0^\dagger \gamma_4 \Lambda^{+\prime}) \\ &= \operatorname{tr}\left(0 \frac{m - i\gamma p}{2m}\, \gamma_4 0^\dagger \gamma_4\, \frac{m' - i\gamma p'}{2m'}\right). \end{aligned} \tag{3.5.44}$$

Some properties of the traces of $\gamma$-matrices are given in Appendix A6.

### 3.5(i). Spin-one systems

Consider a vector wave function

$$\boldsymbol{\psi} = \mathbf{e}\, e^{ikx}$$

which transforms under spatial rotations in the same way as $\mathbf{x}$ (compare § 3.3)

$$\begin{aligned} \psi_i(\mathbf{x}) \to \psi_i'(\mathbf{x}') &= a_{ij} \psi_j(\mathbf{x}) \\ &= (\delta_{ij} - \varepsilon_{ij})\, \psi_j(\mathbf{x}) \\ i, j, k &\equiv 1, 2, 3. \end{aligned} \tag{3.5.45}$$

At the same time we can write [compare §§ 3.3 and 3.5(d)]

$$\begin{aligned} \psi_i'(\mathbf{x}') &= U_{ij} \psi_j(\mathbf{x}) \\ &= \left(\delta_{ij} - \frac{i}{2} \varepsilon_{kl} S_{kl}^{ij}\right) \psi_j(\mathbf{x}) \end{aligned} \tag{3.5.46}$$

hence

$$\varepsilon_{ij} = \frac{i}{2} \varepsilon_{kl} S_{kl}^{ij}$$

which can be satisfied by

$$S_{kl}^{ij} = (S_{kl})_{ij} = -i(\delta_{ki}\, \delta_{lj} - \delta_{kj}\, \delta_{li}) \tag{3.5.47}$$

($ij$ are the matrix elements of $S_{kl}$). If we proceed as in the case of the Dirac equations [§ 3.5(d)] by writing

$$S_1 = S_{23} \qquad S_2 = S_{31} \qquad S_3 = S_{12}$$

then the operators $S$ can be evaluated giving

$$\mathbf{S}^2 = S_1^2 + S_2^2 + S_3^2 = \begin{pmatrix} 2 & 0 & 0 \\ 0 & 2 & 0 \\ 0 & 0 & 2 \end{pmatrix} \equiv s(s+1)\,\hat{1} \tag{3.5.48}$$

and since $2 = 1(1 + 1)$ we have a system for describing spin-one particles. The polarisation vectors $\mathbf{e}$ have a Cartesian base

$$e_1 = \begin{pmatrix} 1 \\ 0 \\ 0 \end{pmatrix} \qquad e_2 = \begin{pmatrix} 0 \\ 1 \\ 0 \end{pmatrix} \qquad e_3 = \begin{pmatrix} 0 \\ 0 \\ 1 \end{pmatrix}$$

which can be diagonalised to

$$e_+ = -\frac{(e_1 + ie_2)}{\sqrt{2}} \qquad e_0 = e_3 \qquad e_- = \frac{(e_1 - ie_2)}{\sqrt{2}} \tag{3.5.49}$$

so that ($S'$ is the diagonalised operator)

$$S'_z\, e_\pm = \pm e_\pm \qquad S'_z\, e_0 = 0. \tag{3.5.50}$$

This is all very fine, but what we want is a four-vector and not a three-vector

$$\psi \to \phi_\mu \qquad \mu = 1, 2, 3, 4$$

$$\phi_\mu = e_\mu e^{ikx}. \tag{3.5.51}$$

We then have the problem of dealing with the fourth component. Since $\phi_\mu$ will behave like $x_\mu$, then $\phi = \boldsymbol{\phi}, \phi_4$ and under spatial notations $\boldsymbol{\phi}$ behaves like a vector as before whilst $\phi_4$ behaves like a scalar, so that we have a mixture of spin 1 and 0 systems. This behaviour is reflected in $S_1^2 + S_2^2 + S_3^2$ which becomes

$$S_1^2 + S_2^2 + S_3^2 = \begin{pmatrix} 2 & 0 & 0 & 0 \\ 0 & 2 & 0 & 0 \\ 0 & 0 & 2 & 0 \\ 0 & 0 & 0 & 0 \end{pmatrix}. \tag{3.5.52}$$

The fourth component is thus an embarrassment and especially in the rest frame of the particle, where we should like our system to have three components only. The simplest way of achieving this is to postulate

$$\frac{\partial \phi_\mu}{\partial x_\mu} = 0 \longrightarrow e_\mu k_\mu = 0 \tag{3.5.53}$$

so that

$$e_4 = -\frac{\mathbf{e}\cdot\mathbf{k}}{k_4} \xrightarrow{\mathbf{k}\to 0} 0.$$

We then have a system with 3 degrees of freedom. The next point we must consider is that $\phi_\mu$ is linked with the reference frame, whereas in general one often wants to relate the polarisation vectors to the direction of say, the particle's motion. We therefore introduce three new vectors $e^i_\mu$, where the symbols $i = 1, 2, 3$ represent a set of orthonormal directions with respect to $\mathbf{k}$, the particle's momentum. The terms $e^i_\mu$ satisfy the normalisation requirement

$$\sum_\mu e^i_\mu e^j_\mu = \delta_{ij} \tag{3.5.54}$$

(i.e. one particle in a box).

The sum over the complete set of spin states then gives

$$\sum_{i=1}^{3} e^i_\mu e^i_\nu = a\,\delta_{\mu\nu} + bk_\mu k_\nu.$$

This is the most general rank two tensor which we can form, since $k$ is the only vector available for making this construction. This expression must satisfy two further requirements

$$\sum_\mu k_\mu e^i_\mu e^i_\nu = 0 \qquad \sum_\nu k_\nu e^i_\mu e^i_\nu = 0$$

and from (3.5.54)

$$\sum_{\mu, i} e^i_\mu e^i_\mu = 3$$

which leads to the solution $a = 1$, $b = 1/m^2$, hence

$$\sum_{i=1}^{3} e^i_\mu e^i_\nu = \delta_{\mu\nu} + \frac{1}{m^2} k_\mu k_\nu. \tag{3.5.55}$$

Now consider photons. At first sight the above summation has the alarming property that $\sum_i \to \infty$ as $m \to 0$. This should not be worrying on two counts:

1. conventional spin has a well-defined meaning only in the rest frame;
2. in the limit $m \to 0$ we cannot expect to be able to work with systems with spatial components only, i.e. $i$ must run $1 \to 4$.

Let us return to the relation (3.5.53)

$$e_\mu k_\mu = 0$$

and note that $e_\mu$ is not well defined by this relation since we find that in the limit $m \to 0$ we can substitute $e_\mu + ck_\mu$ without disturbing the above equation

$$(e_\mu + ck_\mu)\, k_\mu = 0 \qquad m \to 0. \tag{3.5.56}$$

In this equation $c$ is an arbitrary term and so we shall define it as

$$c = -\frac{e_4}{k_4}\,\delta_{\mu 4}$$

we then have

$$e_\mu k_\mu \to \mathbf{e}\cdot\mathbf{k} = 0. \tag{3.5.57}$$

This condition is called the Coulomb gauge condition; the requirement $e_\mu k_\mu = 0$ is known as the Lorentz gauge or subsidiary condition.

The condition

$$\mathbf{e}\cdot\mathbf{k} = 0$$

implies that the polarisation lies at right angles to $\mathbf{k}$. It can be defined completely by two orthogonal vectors, $\mathbf{e}^1$ and $\mathbf{e}^2$ where (Fig. 3.3)

$$\mathbf{e}^1\cdot\mathbf{e}^2 = \mathbf{e}^1\cdot\mathbf{k} = \mathbf{e}^2\cdot\mathbf{k} = 0.$$

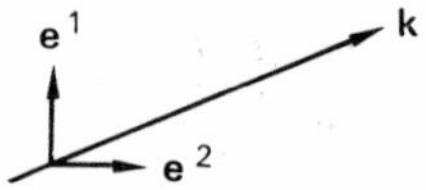

FIG. 3.3

Thus we have a two-component polarisation vector to describe the system; this can only yield independent two spin states. They can be chosen as

$$e_R = -\frac{(\mathbf{e}^1 + i\mathbf{e}^2)}{\sqrt{2}} \qquad e_L = \frac{(\mathbf{e}^1 - i\mathbf{e}^2)}{\sqrt{2}}. \tag{3.5.58}$$

The above relations imply 90° and 270° phase relations between $\mathbf{e}^1$ and $\mathbf{e}^2$. This is always possible since they refer to two independent equations. In optical terms one would speak of right ($R$) or left ($L$) handed circularly polarised light; in the same terms a zero phase relation between $\mathbf{e}^1$ and $\mathbf{e}^2$ implies plane polarised light.

Although only two polarisation vectors are necessary for massless systems it is often formally convenient to add on a further space and a time component. This is always done in such a way that they effectively cancel each other out when observable quantities are calculated. We then have the orthonormality condition

$$\sum_{\lambda=1}^{4} e_\mu^\lambda e_\nu^\lambda = \delta_{\mu\nu}. \tag{3.5.59}$$

## 3.6. Relativistic Fields

The relativistic wave equations we have discussed essentially deal with one-particle systems. In high-energy physics particles are both created and destroyed. Provision for these processes must be included in the theory, which is best approached at first from the concept of fields, and second quantisation. Consider the wave functions $\phi_\mu(x)$, $\psi(x)$ of § 3.5; they already have a space-time dependence and so we can Fourier analyse them as

$$\begin{aligned} \phi_\mu(x) &= \frac{1}{\sqrt{V}} \sum_{\mathbf{k}} N[a_{\mathbf{k}} e^{ikx} + b_{\mathbf{k}}^\dagger e^{-ikx}]\, e_\mu \\ \psi(x) &= \frac{1}{\sqrt{V}} \sum_{\mathbf{p},r} N\, [a_{r\mathbf{p}} u_r e^{ipx} + b_{r\mathbf{p}}^\dagger v_r e^{-ipx}] \end{aligned} \tag{3.6.1}$$

where the terms $N$ are normalisation factors. The Hermitian conjugates of these expressions are $\phi_\mu^\dagger$, $\psi^\dagger$. For the Dirac system a more useful expression is $\bar{\psi} = \psi^\dagger \gamma_4$.

In the process of second quantisation these wave functions are essentially relabelled as fields and at the same time are converted to operators by defining the *a* and *b* terms to have the properties of operators:

$$\text{wave functions} \xrightarrow{\text{second quantisation}} \text{field operators.}$$

Consider a normalised state $|n_{\mathbf{k}}\rangle$ containing $n$ bosons with momentum $\mathbf{k}$

$$\langle n'_{\mathbf{k}'} \mid n_{\mathbf{k}}\rangle = \delta_{n'n}\,\delta_{\mathbf{k}'\mathbf{k}}.$$

Then $a^\dagger$ and $a$ are defined to have the following properties:

$$\begin{aligned} a^\dagger_{\mathbf{k}}\,|n_{\mathbf{k}}\rangle &= \sqrt{(n+1)_{\mathbf{k}}}\,|(n+1)_{\mathbf{k}}\rangle \\ a_{\mathbf{k}}\,|n_{\mathbf{k}}\rangle &= \sqrt{n_{\mathbf{k}}}\,|(n-1)_{\mathbf{k}}\rangle \end{aligned} \tag{3.6.2}$$

thus $a^\dagger$ and $a$ are respectively creation and destruction operators $-a^\dagger$ creates an additional particle of momentum $\mathbf{k}$, whilst $a$ destroys one. Now consider combinations of the operators

$$a_{\mathbf{k}}a^\dagger_{\mathbf{k}}\,|n_{\mathbf{k}}\rangle = (n+1)_{\mathbf{k}}\,|n_{\mathbf{k}}\rangle$$

$$a^\dagger_{\mathbf{k}}a_{\mathbf{k}}\,|n_{\mathbf{k}}\rangle = n_{\mathbf{k}}\,|n_{\mathbf{k}}\rangle$$

but for $\mathbf{k} \neq \mathbf{k}'$

$$a_{\mathbf{k}}a^\dagger_{\mathbf{k}'}\,|n_{\mathbf{k}};n_{\mathbf{k}'}\rangle = \sqrt{n_{\mathbf{k}}}\sqrt{(n+1)_{\mathbf{k}'}}\,|(n-1)_{\mathbf{k}};(n+1)_{\mathbf{k}'}\rangle$$

$$a^\dagger_{\mathbf{k}'}a_{\mathbf{k}}\,|n_{\mathbf{k}};n_{\mathbf{k}'}\rangle = \sqrt{(n+1)_{\mathbf{k}'}}\sqrt{n_{\mathbf{k}}}\,|(n-1)_{\mathbf{k}};(n+1)_{\mathbf{k}'}\rangle$$

therefore $a^\dagger$ and $a$ satisfy the commutation relations

$$[a_{\mathbf{k}}, a^\dagger_{\mathbf{k}'}] = \delta_{\mathbf{k}\mathbf{k}'}. \tag{3.6.3}$$

We also can give similar properties to $b$, so that

$$[b_{\mathbf{k}}, b^\dagger_{\mathbf{k}'}] = \delta_{\mathbf{k}\mathbf{k}'}.$$

We next introduce the occupation number operators

$$N_{\mathbf{k}} = a^\dagger_{\mathbf{k}}a_{\mathbf{k}} \qquad \bar{N}_{\mathbf{k}} = b^\dagger_{\mathbf{k}}b_{\mathbf{k}} \tag{3.6.4}$$

then the eigenvalues of $N_{\mathbf{k}}$ yield the number of particles with momentum $\mathbf{k}$

$$N_{\mathbf{k}}\,|n_{\mathbf{k}}\rangle = n_{\mathbf{k}}\,|n_{\mathbf{k}}\rangle.$$

Since $N$ is the product of an operator times its Hermitian conjugate it is positive definite and so the lowest eigenvalue it can have is zero. Hence a state must exist with the property

$$a_{\mathbf{k}}\,|0\rangle = 0$$

so that

$$N_{\mathbf{k}}\,|0\rangle = 0.$$

This state is called the vacuum state. The one-particle state can then be represented as

$$|\mathbf{k}\rangle = a_{\mathbf{k}}^{\dagger}\,|0\rangle. \qquad (3.6.5)$$

If the normalisation term $N$ in (3.6.1) for the boson fields is chosen to be $1/\sqrt{2\omega_{\mathbf{k}}}$ and the properties of the field equations are exploited, the following expressions are obtained for four momenta and electric currents for the fields [a detailed derivation is given in *P.E.P.*, § 4.3(f) and § 4.3(i)]

$$P_{\lambda} = \sum_{\mathbf{k}} (N_{\mathbf{k}} + \bar{N}_{\mathbf{k}} + \delta_{\lambda 4})\, k_{\lambda} \qquad (3.6.6)$$

$$J_{\lambda} = e \int d\mathbf{x}\, j_{\lambda} = e \sum_{\mathbf{k}} (N_{\mathbf{k}} - \bar{N}_{\mathbf{k}}) \frac{k_{\lambda}}{\omega_{\mathbf{k}}}.$$

These expressions make physical sense if $N$ and $\bar{N}$ represent occupation numbers for particles and antiparticles respectively. The $\delta_{\lambda 4}$ is the zero point energy term and corresponds to the $\frac{1}{2}$ in the energy term for a harmonic oscillator $E = \sum_{k}(n + \frac{1}{2})\omega_{k}$. Note that if we had wished to describe systems with zero charge, then

$$\bar{N}_{k} = N_{k}, \quad \text{i.e.} \quad a_{k} = b_{k}$$

so that

$$\phi = \phi^{\dagger}.$$

In this situation, when we have inserted the normalisation factor, the first of equations (3.6.1) becomes

$$\phi_{\mu}(x) = \frac{1}{\sqrt{V}} \sum_{\mathbf{k}} \frac{1}{\sqrt{2\omega_{\mathbf{k}}}} [a_{\mathbf{k}} e^{ikx} + a_{\mathbf{k}}^{\dagger} e^{-ikx}]\, e_{\mu}. \qquad (3.6.7)$$

Such a situation occurs in fields for $\pi^0$ mesons or photons; in the former case we would set $e_{\mu} = 1$ since the pion is a spinless particle.

If one tries to proceed in a similar fashion for fermions, one expects similar expressions for $P_\lambda$ and $J_\lambda$. This is almost true if one chooses $N = \sqrt{m/E_\mathbf{p}}$, but the signs are wrong

$$P_\lambda = \sum_\alpha (N_\alpha - \overline{N}_\alpha - \delta_{\lambda 4})\, p_\lambda \qquad \alpha \equiv r, \mathbf{p}$$

$$J_\lambda = e \sum_\alpha (N_\alpha + \overline{N}_\alpha) \frac{p_\lambda}{E_p}.$$

These equations carry with them the implications of the existence of negative energies and currents flowing in wrong directions. Sensible equations [i.e. (3.6.6)] can be obtained, however, if one uses anticommutation rather than commutation relations in their derivations

$$a_\alpha a_\beta^\dagger + a_\beta^\dagger a_\alpha = \delta_{\alpha\beta} \equiv \delta_{rr'}\, \delta_{\mathbf{pp}'} \tag{3.6.8}$$

$$a_\alpha^\dagger a_\alpha^\dagger = a_\alpha a_\alpha = 0.$$

These conditions automatically lead to the Pauli exclusion principle for Dirac particles, since if

$$N_\alpha = a_\alpha^\dagger a_\alpha$$

then

$$N_\alpha^2 = a_\alpha^\dagger a_\alpha a_\alpha^\dagger a_\alpha = a_\alpha^\dagger (1 - a_\alpha^\dagger a_\alpha)\, a_\alpha = a_\alpha^\dagger a_\alpha = N_\alpha. \tag{3.6.9}$$

Thus the eigenvalues of $N$ can only be 0, 1 which is the Pauli exclusion principle.

CHAPTER 4

# THE INTERACTION AMPLITUDE

## 4.1. The Perturbation Approach

### 4.1(a). Introduction

One of the most familiar expressions in quantum mechanics is the Hamiltonian equation

$$i\frac{\partial}{\partial t}|\psi\rangle = H|\psi\rangle$$

where $H$ is the total Hamiltonian for any system and contains both the static and interacting parts. Let us assume‡ that we can split off the interacting part and rewrite our equation as

$$i\frac{\partial}{\partial t}|\psi_I\rangle = H_I|\psi_I\rangle \tag{4.1.1}$$

where the subscript $I$ refers to the interaction.

Now consider how we should describe an interaction. We will set up a time scale such that the initial particles start from $t = -\infty$, interact at greatest strength around $t = 0$ and then the final particles go off towards $t = +\infty$. At $t = \pm\infty$ the particles are separated by large distances and are no longer interacting with each other and correspond to the initial $|i\rangle$ and final $|f\rangle$ states discussed in Chapter 2.

‡ We are considerably simplifying an argument here which is developed at some length in §§ 8.1 and 8.2 of *P.E.P.*

Now let us assume that at any time $t$ and $t'$ the states are linked by an operator $U(t, t')$

$$|\psi_I(t)\rangle = U(t, t')\,|\psi_I(t')\rangle \tag{4.1.2}$$

then for $t \to +\infty$, $t' \to -\infty$, $U$ will correspond to the $S$ operator discussed in § 2.2(a)

$$S = \lim_{\substack{t\to\infty \\ t'\to-\infty}} U(t, t') \tag{4.1.3}$$

$$\sum_f |f\rangle = S\,|i\rangle.$$

Now equation (4.1.1) can be regarded as describing the temporal development of $|\psi_I\rangle$ in the time interval $\delta t \to 0$ at any time, and so we can expand with the aid of (4.1.2) and write

$$i\frac{\partial}{\partial t}|\psi_I(t)\rangle = i\frac{\partial}{\partial t}U\,|\psi_I(t')\rangle = H_I\,|\psi_I(t)\rangle = H_I U\,|\psi_I(t')\rangle.$$

We thus obtain an operator equation

$$i\frac{\partial U}{\partial t} = H_I U \tag{4.1.4}$$

and we wish to seek solutions for it. So far our treatment has been exact. We now introduce the perturbation approach by assuming that the changes induced by $H_I$ in the interval $\delta t$ are small, and so we can expand $U$ in ascending powers of $H_I$

$$U = \hat{1} + U_1 + U_2 + U_3 + \cdots \tag{4.1.5}$$

where $U_1$ is of first order in $H_I$ and so on. Thus if we substitute this equation in (4.1.4) and equate terms of equal order we obtain

$$i\frac{\partial U_1}{\partial t} = H_I$$

$$i\frac{\partial U_2}{\partial t} = H_I U_1 \tag{4.1.6}$$

$$\cdots\cdots\cdots$$

These equations possess the solutions

$$U_1(t, t') = -i \int_{t'}^{t} dt_1\, H_I(t_1)$$

$$U_2(t, t') = (-i)^2 \int_{t'}^{t} dt_1 \int_{t'}^{t_1} dt_2\, H_I(t_1)\, H_I(t_2)$$

$$\cdots\cdots\cdots\cdots$$

so that

$$U = \hat{1} - i \int_{t'}^{t} dt_1\, H_I(t_1) + (-i)^2 \int_{t'}^{t} dt_1 \int_{t'}^{t_1} dt_2\, H_I(t_1)\, H_I(t_2) + \cdots$$

Now since the interaction spreads in space as well as time we can introduce a Hamiltonian density $\mathscr{H}_I$

$$\int dt\, H_I(t) = \int dt\, d\mathbf{x}\, \mathscr{H}_I(\mathbf{x}, t) = \int d^4x\, \mathscr{H}_I(x)$$

and so in the limit $t \to \infty$, $t' \to -\infty$

$$S \equiv U = \hat{1} - i \int_{-\infty}^{+\infty} d^4x_1\, \mathscr{H}_I(x_1)$$

$$+ \frac{(-i)^2}{2!} \int_{-\infty}^{+\infty} d^4x_1\, d^4x_2\, \mathscr{H}_I(x_1)\, \mathscr{H}_I(x_2) + \cdots \qquad (4.1.7)$$

where the insertion of $+\infty$ for the upper limit on $x_2$ and the factor 2! is based on an argument due to Dyson (1949) which allows us to write terms in this form providing the earliest operators in time appear on the right (*P.E.P.*, p. 295).

### 4.1(b). Application to rho-meson decay

As an example of the perturbation technique consider the decay of the $\varrho$-meson (Fig. 4.1).

The Hamiltonian for the interaction must be Lorentz invariant, i.e. a scalar under Lorentz transformations. Since the $\varrho$-meson is a spin one particle the associated field $\phi_\mu$ is a four vector [§ 3.5(i)]. The pion is a

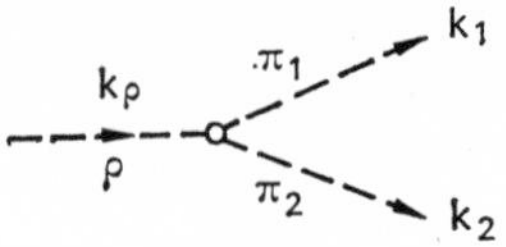


FIG. 4.1

spinless particle and its associated field is a scalar (more strictly pseudoscalar since spin-parity for the pion is $0^-$). We must therefore produce another four vector to combine with $\phi_\mu$ to form a scalar product; this can be done by representing one pion field by $\partial\phi/\partial x_\mu$. Thus the simplest form the Hamiltonian can take is

$$\mathscr{H}_I = i\gamma\left(\phi_1\frac{\partial\phi_2}{\partial x_\mu} \pm \phi_2\frac{\partial\phi_1}{\partial x_\mu}\right)\phi_\mu \tag{4.1.8}$$

where $\gamma$ is a coupling strength and we shall ignore complexities like isospin for the particles. The significance of $\pm$ will emerge later.

The $S$ matrix expansion (4.1.7) will be stopped at the lowest possible term which allows a sensible calculation. This gives us

$$\begin{aligned} S_{fi} &= \langle k_1k_2|\,\hat{1}\,|k_\varrho\rangle + \langle k_1k_2|\,U_1\,|k_\varrho\rangle + \langle k_1k_2|\,U_2\,|k_\varrho\rangle + \cdots \\ &= -i\langle k_1k_2|\int d^4x\,\mathscr{H}_I\,|k_\varrho\rangle \end{aligned} \tag{4.1.9}$$

where the first term vanished since the states are orthogonal.

In the present circumstances the field operator given in (3.6.7) is adequate for both rho and pion fields; in the latter case we of course set $e_\mu = 1$ and reduce $\phi$ to a one-component field. The action of $\phi_\mu$ on the state $|k_\varrho\rangle$ yields, with the aid of the commutation relation (3.6.3)

$$\begin{aligned} \phi_\mu(x)\,|k_\varrho\rangle &= \frac{1}{\sqrt{V}}\sum_{\mathbf{k}}\frac{e_\mu}{\sqrt{2\omega_{\mathbf{k}}}}\,[a_{\mathbf{k}}e^{ikx} + a^\dagger_{\mathbf{k}}e^{-ikx}]\,a^\dagger_{\mathbf{k}_\varrho}\,|0\rangle \\ &= \frac{1}{\sqrt{V}}\sum_{\mathbf{k}}\frac{e_\mu e^{ikx}}{\sqrt{2\omega_{\mathbf{k}}}}\,[\delta_{\mathbf{k}\mathbf{k}_\varrho} + a^\dagger_{\mathbf{k}_\varrho}a_{\mathbf{k}}]\,|0\rangle + \frac{1}{\sqrt{V}}\sum_{\mathbf{k}}\frac{e_\mu e^{-ikx}}{\sqrt{2\omega_{\mathbf{k}}}}\,|k_\varrho k\rangle \\ &= \frac{1}{\sqrt{V}}\,\frac{e_\mu}{\sqrt{2\omega_\varrho}}\,e^{ik_\varrho x}\,|0\rangle + \frac{1}{\sqrt{V}}\sum_{\mathbf{k}}\frac{e_\mu}{\sqrt{2\omega_{\mathbf{k}}}}\,e^{-ikx}\,|k_\varrho k\rangle. \end{aligned}$$

Similarly the pion field operators give

$$\langle k_1 | \varphi_1(x) = \frac{1}{\sqrt{V}} \langle 0 | \frac{e^{-ik_1x}}{\sqrt{2\omega_1}} + \frac{1}{\sqrt{V}} \sum_{\mathbf{k}_1'} \langle k_1 k_1' | \frac{e^{ik_1'x}}{\sqrt{2\omega_1'}}$$

$$\langle k_2 | \varphi_2(x) = \frac{1}{\sqrt{V}} \langle 0 | \frac{e^{-ik_2x}}{\sqrt{2\omega_2}} + \frac{1}{\sqrt{V}} \sum_{\mathbf{k}_2'} \langle k_2 k_2' | \frac{e^{ik_2'x}}{\sqrt{2\omega_2'}}.$$

Thus we have to combine eight states. In fact only the first term in each expansion contributes as the remaining states are all orthogonal to each other. Therefore, we are reduced to one term; including the differentials we find

$$S_{fi} = i\langle 0 \mid 0\rangle \int d^4x\, i\gamma\, ie_\mu (k_2 \pm k_1)_\mu \frac{1}{V^{3/2}} \frac{1}{\sqrt{8\omega_1\omega_2\omega_\varrho}} e^{i(k_\varrho - k_1 - k_2)x}. \tag{4.1.10}$$

The significance of the $\pm$ alternatives now emerges. Both choices lead to four vectors, but the properties of spin one equations (3.5.53) give

$$e_\mu (k_1 + k_2)_\mu = e_\mu k_{\varrho\mu} = 0.$$

Hence

$$S_{fi} = \frac{1}{V^{3/2}} \frac{1}{\sqrt{8\omega_1\omega_2\omega_\varrho}} i\gamma\, e_\mu (k_1 - k_2)_\mu \int d^4x\, e^{i(k_\varrho - k_1 - k_2)x}$$

$$= i(2\pi)^4 \frac{1}{V^{3/2}} \frac{1}{\sqrt{8\omega_1\omega_2\omega_\varrho}} \gamma\, e_\mu (k_1 - k_2)_\mu\, \delta(k_\varrho - k_1 - k_2).$$

If we recall our original expansions of the $S$ matrix [(2.2.11), (2.2.13)]

$$S_{fi} = \delta_{fi} + i(2\pi)^4\, \delta(p_i - p_f)\, M_{fi}$$

$$= \delta_{fi} + i(2\pi)^4\, \delta(p_i - p_f) \frac{1}{N} T_{fi}$$

then the matrix elements $M_{fi}$, $T_{fi}$ are easily recognisable

$$M_{fi} = \frac{1}{V^{3/2}} \frac{1}{\sqrt{8\omega_1\omega_2\omega_\varrho}} \gamma\, e_\mu (k_1 - k_2)_\mu \tag{4.1.11}$$

$$T_{fi} = \gamma\, e_\mu (k_1 - k_2)_\mu . \tag{4.1.12}$$

The decay rate calculated from this amplitude is given in equation (5.3.7).

## 4.2. Invariance Arguments

Once one understands the basic action of the field operators, the same results for $T_{fi}$ can often be obtained more quickly and simply by invariance arguments. In the case of

$$\varrho \to \pi\pi$$

one needs a polarisation vector $e_\mu$ for the $\varrho$ and since we are not requiring spin (polarisation) dependent terms then $T_{fi}$ must be a scalar; now the only four vectors we have to combine with $e_\mu$ are $k_{1\mu}$ and $k_{2\mu}$ for the pions hence

$$\begin{aligned} T_{fi} &= \gamma\, e_\mu(k_1 - k_2)_\mu + \gamma'\, e_\mu(k_1 + k_2)_\mu \\ &= \gamma\, e_\mu(k_1 - k_2)_\mu \end{aligned} \tag{4.2.1}$$

since $e_\mu k_{\varrho\mu} = 0$. The term $\gamma$ must be a relativistically invariant function of $k_\varrho$, $k_1$ and $k_2$ and therefore constant in the present situation since $k_\varrho = k_1 + k_2$ forces it to depend on the masses involved. Since $e_\mu(k_1 - k_2)_\mu$ contains all the invariance properties of the amplitude $\gamma$ must contain all the dynamical contributions; in effect it represents $U_1 + U_2 + U_3 \ldots$ in the perturbation expansion when the polarisation and kinematic terms have been factored out. Thus there is a difference in emphasis on the role of $\gamma$ in the perturbation and invariance approaches. In the former it is regarded as a constant and the next higher term in the perturbation expansion would involve $\gamma^2$, whilst in the latter its formulation as a scalar function of the relevant kinematic variables implies that it does not necessarily have a constant value in all situations.

Now consider a more complicated case

$$\varrho \to \pi\gamma .$$

The relevant variables are

| | $\varrho$ | $\pi$ | $\gamma$ |
|---|---|---|---|
| four momenta | $k_\mu$ | $k_{\pi\mu}$ | $k'_\mu$ |
| polarisation | $e_\mu$ | 1 | $e'_\mu$. |

These satisfy the conditions

$$k = k_\pi + k' \qquad e_\mu k_\mu = e'_\mu k'_\mu = 0,$$

hence one $k$ variable is redundant, nor can we use terms of type $k \cdot k'$ since they can be expressed in terms of masses $[-m_\pi^2 = (k - k')^2]$. Possible candidates are then

$$T_{fi} = \alpha\, e_\mu e'_\mu + \beta(e_\mu k'_\mu)\,(e'_\nu k_\nu) + \gamma \varepsilon_{\mu\nu\lambda\sigma}\, e_\mu e'_\nu k_\lambda k'_\sigma.$$

The ambiguity in $e'_\nu$ for massless systems (§ 3.5(i)) leads to the gauge invariance requirement on the substitution $e'_\nu \to e'_\nu + ck'_\nu$. The first two terms fail this requirement, hence $\alpha = \beta = 0$, and

$$T_{fi} = \gamma \varepsilon_{\mu\nu\lambda\sigma}\, e_\mu e'_\nu k_\lambda k'_\sigma. \tag{4.2.2}$$

where $\gamma$ is a Lorentz invariant function of the dynamical variables (in fact $\gamma$ is a constant in two-body decay since only masses are involved).

## 4.3. Propagators

Consider the scattering of two particles which we shall label as in Fig. 4.2 (the $k$s are the four momenta).

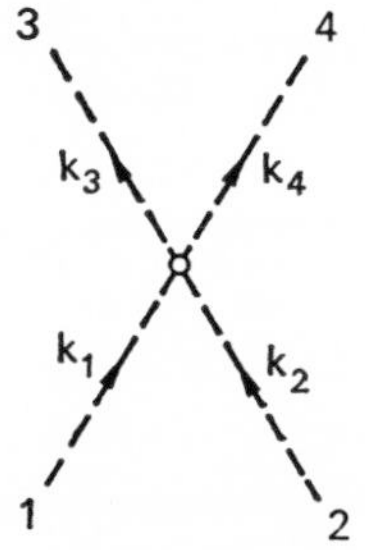


FIG. 4.2

The associated amplitudes are frequently expressed in terms of the Lorentz invariant parameters $s$, $t$ and $u$, defined as

$$\begin{aligned} s &= -(k_1 + k_2)^2 \\ t &= -(k_1 - k_3)^2 \\ u &= -(k_1 - k_4)^2 \end{aligned} \tag{4.3.1}$$

In the $c$-system $s$ gives the square of the total energy

$$s = -(\mathbf{k}_1 + \mathbf{k}_2)^2 + (E_1 + E_2)^2 = E_c^2 \quad \text{since by definition} \quad \mathbf{k}_1 = -\mathbf{k}_2 \tag{4.3.2}$$

whilst $t$ and $u$ represent squares of four momentum transfers in any reference frame. It is a straightforward matter to show that

$$s + t + u = m_1^2 + m_2^2 + m_3^2 + m_4^2. \tag{4.3.3}$$

Next let us examine some of the implications of unitarity of the $S$ operator (2.2.4)

$$SS^\dagger = \hat{1}$$

for the transition amplitude. Writing $S = \hat{1} + iR$ we find

$$SS^\dagger = (\hat{1} + iR)(\hat{1} - iR^\dagger) = \hat{1}$$

$$i(R - R^\dagger) = -RR^\dagger$$

$$i(\langle f|\, R\, |i\rangle - \langle f|\, R^\dagger\, |i\rangle) = -\sum_n \langle f|\, R\, |n\rangle \langle n|\, R^\dagger\, |i\rangle \tag{4.3.4}$$

where $\sum_n |n\rangle \langle n|$ represents a complete set of intermediate states. Since equation (2.2.13) gives

$$\langle f|\, R\, |i\rangle = (2\pi)^4\, \delta(p_f - p_i) \frac{1}{N} \langle f|\, T\, |i\rangle$$

and if time reversal holds

$$\langle f|\, T^\dagger\, |i\rangle = \langle i|\, T\, |f\rangle^* = \langle f|\, T\, |i\rangle^*$$

equation (4.3.4) becomes

$$\frac{(2\pi)^4}{N} \delta(p_f - p_i)\, 2\, \mathrm{Im}\, T_{fi} = \sum_n \frac{(2\pi)^8}{NN_n^2} \delta(p_f - p_n)\, \delta(p_n - p_i) \times \langle f|\, T\, |n\rangle \langle n|\, T^\dagger\, |i\rangle$$

where $N_n$ represents the normalisation factor for the intermediate states. The $\delta$-functions can be rewritten as

$$\delta(p_f - p_n)\, \delta(p_n - p_i) = \delta(p_f - p_i)\, \delta(p_n - p_i)$$

and so

$$2\, \mathrm{Im}\, T_{fi} = \sum_n \frac{(2\pi)^4}{N_n^2} \delta(p_n - p_i) \langle f|\, T\, |n\rangle \langle n|\, T^\dagger\, |i\rangle. \tag{4.3.5}$$

Next consider the implications of the sum over the intermediate states. This could involve one particle, two particles or a whole continuum of states. We will make the extreme assumption that the summation is dominated by a single particle intermediate state in the $s$ channel, by this we mean the situation shown in Fig. 4.3.

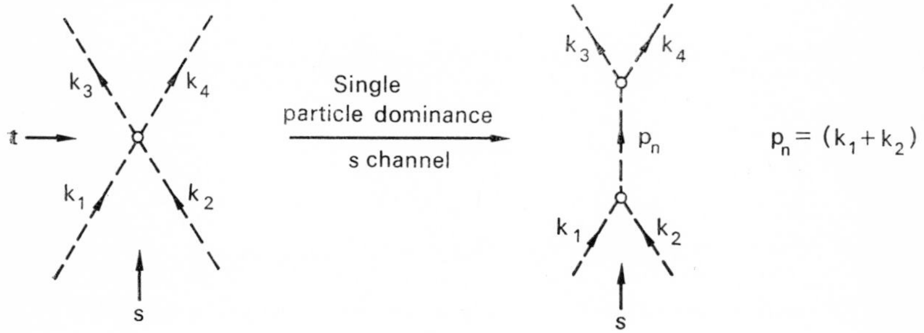


FIG. 4.3

The summation over the momentum states of $n$ gives

$$\sum_n \frac{1}{N_n^2} = \frac{t}{(2\pi)^3} \int \frac{d\mathbf{p}_n}{E_n} = \frac{2t}{(2\pi)^3} \int d^4p_n \, \delta(p_n^2 + m_n^2) \, \vartheta(E_n) \qquad (4.3.6)$$

where [compare (2.3.6)]

$$\begin{aligned} t &= \tfrac{1}{2} \text{ bosons} \qquad & \vartheta(E_n) &= 1 \quad & E_n > 0 \\ &= m \text{ fermions} & &\;\;\, 0 & E_n < 0. \end{aligned}$$

Equation (4.3.6) arises from the properties of the $\delta$-function (Appendix A4)

$$\begin{aligned} \int d^4p\delta(p^2 + m^2)\,\vartheta(E) &= \int d\mathbf{p}\, dE\, \delta(\mathbf{p}^2 + m^2 - E^2)\,\vartheta(E) \\ &= \int d\mathbf{p}\, dE\, \delta(E_p^2 - E^2)\,\vartheta(E) \\ &= \int d\mathbf{p} \frac{dE}{2E} [\delta(E_p - E) + \delta(E_p + E)]\,\vartheta(E) \\ &= \int \frac{d\mathbf{p}}{2E_p}. \end{aligned}$$

Thus, if we use $\sum$ to represent the summation over all states of $|n\rangle$ apart from momentum, Im $T_{fi}$ becomes

$$\begin{aligned}\operatorname{Im} T_{fi} &= \frac{t}{(2\pi)^3}\sum\int d^4p_n\,\delta(p_n^2+m_n^2)\,\vartheta(E_n)\,(2\pi)^4\,\delta(p_n-p_i)\\ &\quad\times\langle f|\,T\,|n\rangle\,\langle n|\,T^\dagger\,|i\rangle\\ &= 2\pi t\sum\langle f|\,T\,|n\rangle\,\langle n|\,T^\dagger\,|i\rangle\,\delta(p_i^2+m_n^2)\\ &= 2\pi t\sum\langle f|\,T\,|n\rangle\,\langle n|\,T^\dagger\,|i\rangle\,\delta(m_n^2-s)\end{aligned}\tag{4.3.7}$$

since by (4.3.1)

$$p_i^2 = (k_1+k_2)^2 = -s.$$

Again we emphasise that this is a summation over a single particle state, and the assumption is made that higher-order terms are negligible.

We also make one more assumption, and this is not approximation, that $T_{fi}$ is an analytic function of $s$. This implies that if the integration of $T_{fi}$ round a closed curve in a complex variable space has the property

$$\int_C ds'\,T_{fi}(s') = 0$$

then Cauchy's theorem‡ leads to the relation

$$T_{fi}(s) = \frac{1}{\pi}\int_{-\infty}^{+\infty} ds'\,\frac{\operatorname{Im} T_{fi}(s')}{s'-s}\tag{4.3.8}$$

hence

$$\begin{aligned}T_{fi}(s) &= \frac{1}{\pi}\int_{-\infty}^{+\infty} ds'\,\frac{2\pi t}{s'-s}\sum\langle f|\,T\,|n\rangle\,\langle n|\,T^\dagger\,|i\rangle\,\delta(m_n^2-s')\\ &= 2t\sum\frac{\langle f|\,T\,|n\rangle\,\langle n|\,T^\dagger\,|i\rangle}{m_n^2-s}.\end{aligned}\tag{4.3.9}$$

‡ Some discussion of the properties of complex variables may be found in § 10.2 of *P.E.P.* A more extensive examination, which at the same time is relevant to particle physics, may be found in Burkhardt (1969).

The expression

$$2t \sum \frac{|n\rangle \langle n|}{m_n^2 - s} \tag{4.3.10}$$

is called the propagator of the interaction. It is of great importance in elementary particle physics, especially in the theory of electromagnetic interactions.

We shall now illustrate the use of equation (4.3.9) by taking a specific example–the scattering process $\pi^+\pi^- \to \pi^+\pi^-$. Since the $\varrho^0$ meson decays to $\pi^+\pi^-$ and the interaction strength $\gamma$ is known to be large we shall postulate that the scattering process proceeds via an intermediate $\varrho$ state (Fig. 4.4).

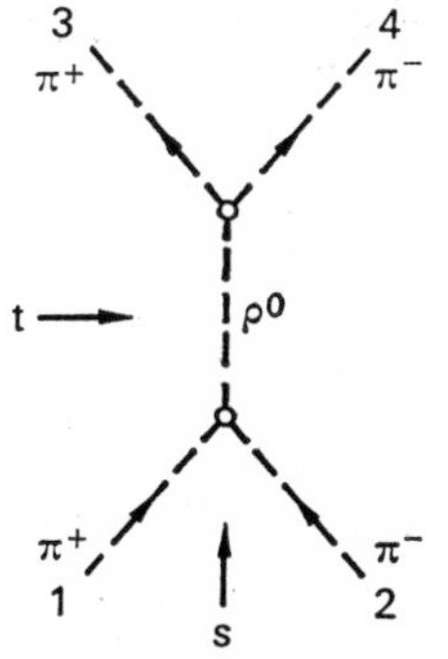


FIG. 4.4

If we again ignore the complications of isospin and take our amplitude from $\varrho$-decay (4.2.1) a suitable form for the interaction amplitude would be

$$T_{fi}(s) = \gamma^2 \sum \frac{(k_3 - k_4)_\mu \, e_\mu e_\nu (k_1 - k_2)_\nu}{[m_\varrho - i(\Gamma/2)]^2 - s}$$

$$= \gamma^2 \frac{(k_3 - k_4)_\mu \left( \delta_{\mu\nu} + \dfrac{k_\mu k_\nu}{m_\varrho^2} \right) (k_1 - k_2)_\nu}{[m_\varrho - i(\Gamma/2)]^2 - s} \tag{4.3.11}$$

where we have used the summation over spin states given in (3.5.55) and allowance has been made for the finite width of the $\varrho$-meson in the denominator. Figure 4.4 and the corresponding amplitude (4.3.11) is called an $s$ channel contribution [recall (4.3.1)]. A further contribution to the amplitude for $\pi^+\pi^-$ scattering can arise in the $t$ channel where a $\varrho^0$ meson can also form the intermediate state (Fig. 4.5); no contribution is possible in the $u$ channel since conservation of charge would require the exchange of a $\varrho^{++}$ meson—this does not exist.

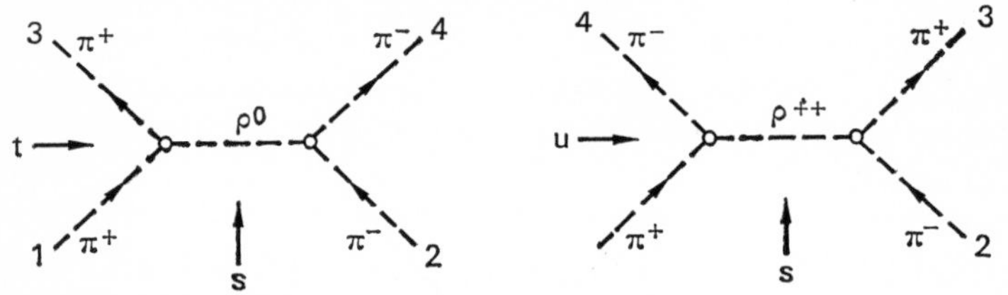


FIG. 4.5

Thus the complete matrix element is

$$T_{fi} = T_{fi}(s) + T_{fi}(t)$$

where the second term is similar in basic form to (4.3.11) with appropriate kinematic alterations‡.

Next consider electromagnetic interactions. Here the basic interaction is the coupling of the photon to an electric current (3.5.14). In the equations below $A_\mu$ is the photon field, $e$ the electric charge and $e_\mu$ the photon polarisation vector [§ 3.5(i)]. In a similar manner to equations (4.1.8) and (4.2.1) we may represent the Hamiltonian and matrix element as

$$\mathscr{H}_I = j_\mu A_\mu = i\,e\,\bar{\psi}\gamma_\mu\psi A_\mu \tag{4.3.12}$$

$$T_{fi} = i\,e\,\bar{u}'\gamma_\mu u\,e_\mu. \tag{4.3.13}$$

This expression is clearly Lorentz invariant, and also satisfies gauge invariance [§ 3.5(i)]—a quick test of gauge invariance is to note that the

‡ An inspection of Fig. 4.5 in conjunction with § 4.2 shows the alterations to be $(k_3 - k_4) \to (k_2 + k_4)$, $(k_1 - k_2) \to (k_1 + k_3)$ and $s \to t$.

substitution of $k_\mu$ for $e_\mu$ should make the expression vanish. Thus from Fig. 4.6

$$\begin{aligned} T_{fi} &= i\, e\, \bar{u}'\gamma_\mu u\, e_\mu \xrightarrow[\text{invariance}]{\text{gauge}} i\, e\, \bar{u}'\gamma_\mu u(p - p')_\mu \\ &= i\, e\, \bar{u}'u(m - m) \\ &= 0 \end{aligned} \tag{4.3.14}$$

since by (3.5.29) and (3.5.11)

$$(i\gamma p + m)\, u = 0 \qquad \bar{u}'(i\gamma p' + m) = 0.$$

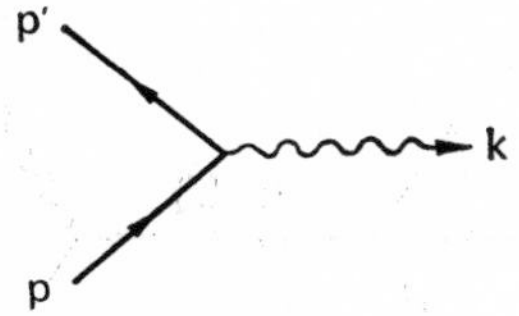


FIG. 4.6

As an example of the application of (4.3.13) consider Compton scattering on electrons. The appropriate matrix element in the $s$ channel (Fig. 4.7) would be [compare (4.3.9)]

$$T_{fi}(s) = e^2\, 2t \sum \bar{u}_{p'} \frac{\gamma_\mu\, e_\mu\, u\bar{u}\gamma_\nu\, e_\nu}{m^2 - s}\, u_p.$$

Now $t = m$ for fermions and by (3.5.43)

$$\sum u\bar{u} = \Lambda^+ = \frac{-i\,\gamma(k + p) + m}{2m}$$

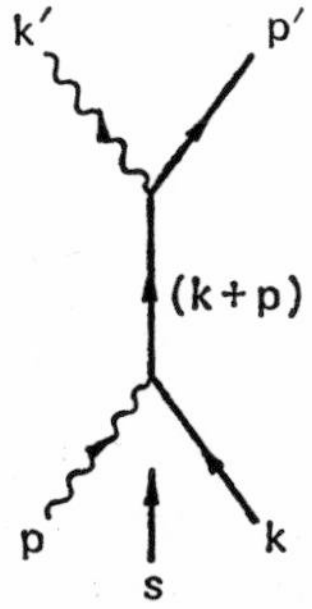


FIG. 4.7

hence

$$T_{fi}(s) = -e^2 \bar{u}_{p'} \gamma_\mu e_\mu \frac{-i\gamma(k+p)+m}{m^2 - s} \gamma_\nu e_\nu u_p .$$

No contribution is possible in the $t$ channel in this case but an additional term arises in the $u$ channel (Fig. 4.8); the appropriate spin projection operator is

$$\sum u\bar{u} = \frac{-i\gamma(p-k')+m}{2m} .$$

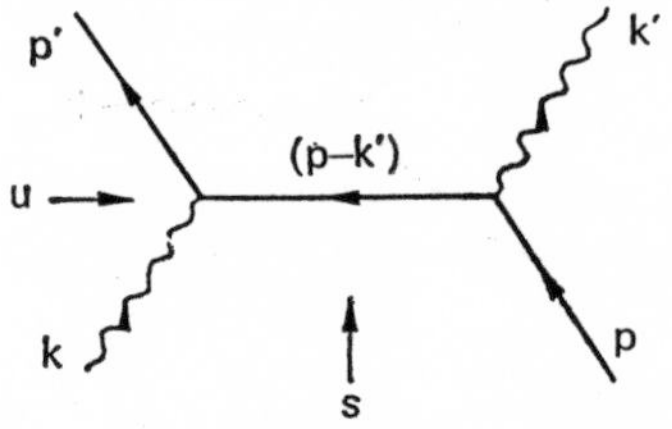


FIG. 4.8

## 4.4. Feynman Rules

It was pointed out by Feynman [1949; see also Dyson (1949)] that a simple one-to-one relation existed between a possible matrix element which could be used to describe an interaction and a diagram which gives it visual interpretation. One can thus draw diagrams for any physical process and then relate the amplitude to them as we have done in the previous sections.

The appropriate rules are summarised in Table 4.1. We follow the Feynman convention that arrows on antifermions are placed in the opposite direction to their four momenta, thus ensuring continuity of current lines. Factors like $(k_1 - k_2)_\mu$ which arose in our expression for $\varrho$-decay do not appear explicitly in the rules; they come under the remark in vertices "factors required to maintain the invariance properties of interactions". The factor $\delta_P = (-1)^n$ arises from the antisymmetry properties of fermions.

TABLE 4.1

FEYNMAN RULES

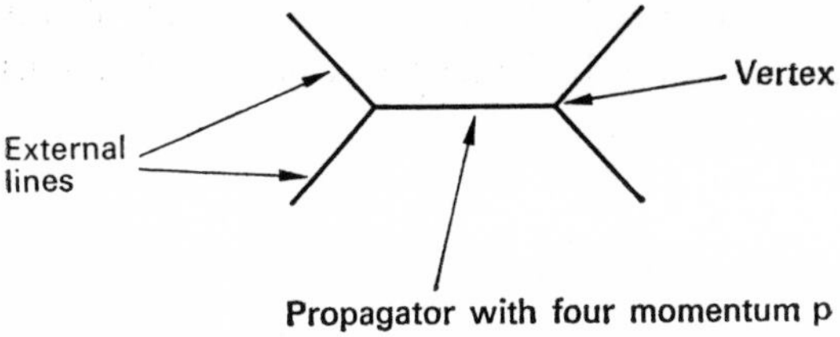


Matrix element
= sum of all possible diagrams

*Propagators*

$$2t.\ \frac{\text{Projection operator for spin state}}{p^2 + m^2} \qquad t = \frac{1}{2}\ \text{bosons},\ m\ \text{fermions}$$

Spin 0 $\dfrac{1}{p^2 + m^2}$ Spin 1 $m \neq 0$ $\dfrac{\delta_{\mu\nu} + \dfrac{p_\mu p_\nu}{m^2}}{p^2 + m^2}$

Spin $\frac{1}{2}$ $\dfrac{-i\gamma p + m}{p^2 + m^2}$ Spin 1 $m = 0$ $\dfrac{\delta_{\mu\nu}}{p^2}$

N.B. Substitution of antiparticle for particle leads to overall factor of $(-1)^{2s}$.

*External lines*

| Particle | spin | in | out |
|---|---|---|---|
| Meson | 0 | $1 \equiv$  | $1 \equiv$  |
| | 1 | $e_\mu \equiv$  | $e_\mu \equiv$  |
| Photon | 1 | $e_\mu \equiv$  | $e_\mu \equiv$  |
| Fermion | $\frac{1}{2}$ | $u \equiv$  | $\bar{u} \equiv$  |
| Antifermion | $\frac{1}{2}$ | $\bar{v} \equiv$  | $v \equiv$  |

*Vertices*

Compounded from coupling strengths and factors required to maintain the invariance properties of interaction. Multiply overall by factor $\delta_P = (-1)^n$ where $n$ is the number of interchanges of pairs of fermions in the final state.

CHAPTER 5

# ELECTROMAGNETIC INTERACTIONS

## 5.1. The Coulomb Scattering of Two Electrons

The basic diagrams representing the scattering are shown in Fig. 5.1.

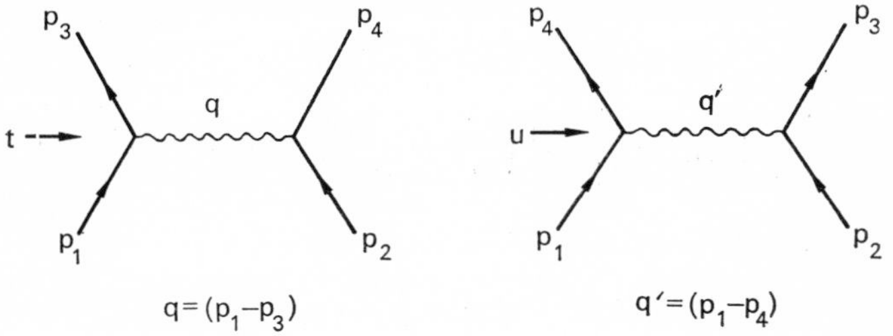


FIG. 5.1

Using the Feynman rules [see also (4.3.13)] the matrix element is

$$T_{fi} = -e^2\bar{u}_3\gamma_\mu u_1 \frac{1}{q^2}\bar{u}_4\gamma_\mu u_2 + e^2\bar{u}_4\gamma_\nu u_1 \frac{1}{q'^2}\bar{u}_3\gamma_\nu u_2. \qquad (5.1.1)$$

Let us consider the non-relativistic limit. Equation (3.5.30) tells us that the *us* can be written as

$$u = \begin{pmatrix}1\\0\\0\\0\end{pmatrix} \quad \text{or} \quad \begin{pmatrix}0\\1\\0\\0\end{pmatrix} \quad \text{compound to } u = \begin{pmatrix}1\\0\end{pmatrix} \qquad (5.1.2)$$

where 1, 0 are 2 × 2 matrices. Use of the $\gamma$ matrices (3.5.4) then gives us

$$\bar{u} = u^{\dagger}\gamma_4 = (1\ \ 0)\begin{pmatrix}1 & 0\\ 0 & -1\end{pmatrix} = (1\ \ 0)$$

$$\bar{u}\gamma_\mu u = \bar{u}\boldsymbol{\gamma} u, \quad \bar{u}\gamma_4 u$$

$$= (1\ \ 0)\begin{pmatrix}0 & -i\boldsymbol{\sigma}\\ i\boldsymbol{\sigma} & 0\end{pmatrix}\begin{pmatrix}1\\ 0\end{pmatrix}, \qquad (1\ \ 0)\begin{pmatrix}1 & 0\\ 0 & -1\end{pmatrix}\begin{pmatrix}1\\ 0\end{pmatrix}$$

$$= 0, 1$$

so that

$$T_{fi} = -e^2\left(\frac{1}{q^2} - \frac{1}{q'^2}\right) = e^2\left(\frac{1}{t} - \frac{1}{u}\right) \tag{5.1.3}$$

where we have used (4.3.1).

We require $\sum_{i,f} |T_{fi}|^2$ and recall that our spinors $\begin{pmatrix}1\\ 0\end{pmatrix}$ each contain two spin states. Altogether we have four initial spin states ↑↑, ↑↓, ↓↑, ↓↓ and since we have no spin flip terms in $T_{fi}$ the final states will be the same, hence

$$\overline{\sum_i}\sum_f |T_{fi}|^2 = \frac{1}{4}4e^4\left(\frac{1}{q^2} - \frac{1}{q'^2}\right)^2 = e^4\left(\frac{1}{t} - \frac{1}{u}\right)^2.$$

If we work in the $c$-system (2.3.9), then

$$q = (\mathbf{p}_1 - \mathbf{p}_3),\ i(E_1 - E_3) = \mathbf{p}_1 - \mathbf{p}_3 \qquad p_c = |\mathbf{p}_1| = |\mathbf{p}_3|$$

$$q^2 = -t = (\mathbf{p}_1 - \mathbf{p}_3)^2 = \mathbf{p}_1^2 - 2\mathbf{p}_1\cdot\mathbf{p}_3 + \mathbf{p}_3^2 = 2p_c^2(1 - \cos\vartheta)$$

$$= 4p_c^2\sin^2\frac{\vartheta}{2}$$

$$q'^2 = -u = (\mathbf{p}_1 - \mathbf{p}_4)^2 = (\mathbf{p}_1 + \mathbf{p}_3)^2 = 4p_c^2\cos^2\frac{\vartheta}{2}.$$

The factors given above can be inserted in our basic expression for the differential cross-section (2.4.3) ($m$ is the electron mass).

$$\frac{d\sigma}{d\Omega} = \frac{1}{(2\pi E_c)^2}\overline{\sum_i}\sum_f |T_{fi}|^2\frac{1}{n^2} \qquad n^2 = \frac{1}{m^4}$$

(4 fermions)

$$E_c = 2m \quad \text{non-relativistic limit}$$

yielding
$$\frac{d\sigma}{d\Omega} = \frac{1}{16\pi^2 m^2}\frac{e^4}{16p_c^4}\left(\frac{1}{\sin^2\vartheta/2} - \frac{1}{\cos^2\vartheta/2}\right)^2 m^4$$
$$= \frac{\alpha^2 m^2}{16p_c^4}\left(\frac{1}{\sin^2\vartheta/2} - \frac{1}{\cos^2\vartheta/2}\right)^2 \qquad \alpha = \frac{e^2}{4\pi} \tag{5.1.4}$$

This is the Rutherford scattering formula. An inspection of the contributing terms shows that the photon exchange in the $t$-channel causes a forward peak (Fig. 5.2) whilst the exchange in the $u$-channel causes a backward peak.

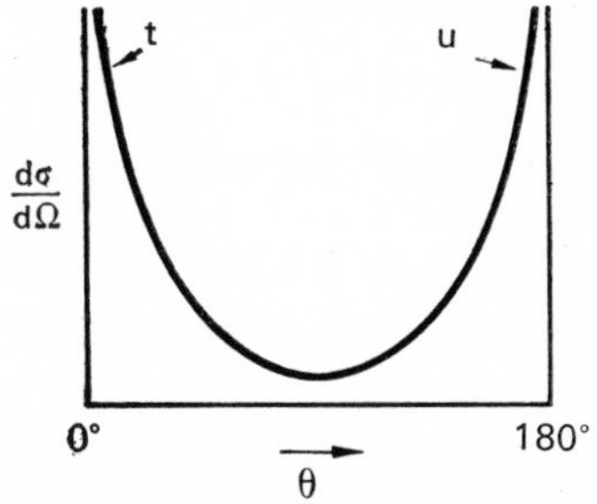


FIG. 5.2. Cross-section for non-relativistic $e^-e^-$ scattering showing the regions of dominance for the $t$- and $u$-channel amplitudes.

Now consider $e^+e^-$ scattering, photon exchange is no longer possible in the $u$-channel (we have no doubly charged photon) and instead we have an $s$-channel contribution (Fig. 5.3). Since in the $c$-system

$$q'^2 = (p_1 + p_2)^2 \to -4m^2$$

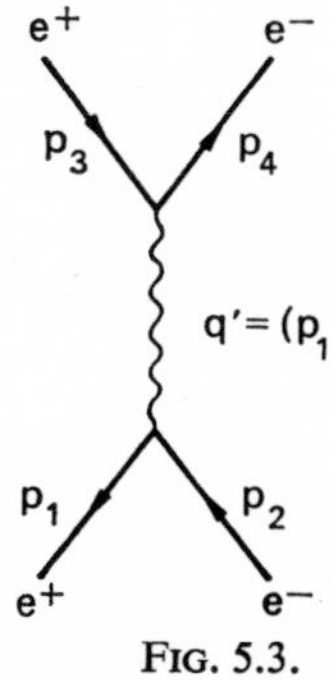


FIG. 5.3.

in the non-relativistic limit, the contribution from this diagram is negligible and so only the forward peak occurs in $e^+e^-$ scattering.

If the scattering is relativistic then the Dirac spin functions become more complicated, but the complications are all in the numerator. The $q^2$, $q'^2$ dependence still holds in the denominators. Experimentally it is found that the two diagrams provide an exact description of the coulomb scattering of two electrons at all energies so far available.

## 5.2. Tests of Quantum Electrodynamics

In fact all the processes involving the coulomb interaction of charged leptons reveal no discrepancies between calculation and experiment. This has led to a strong experimental and theoretical effort to find possible breakdowns in quantum electrodynamics (and also to quantify them if they are found). Supposing in $e^-e^-$ scattering it is possible to exchange a massive vector boson as well as a photon then the denominator of the propagator could be modified to be

$$\frac{1}{q^2} \to \frac{1}{q^2} - \frac{1}{q^2+\Lambda^2} = \frac{1}{q^2}\frac{\Lambda^2}{q^2+\Lambda^2} \sim \frac{1}{q^2}[1-(q^2/\Lambda^2)].$$

Thus departures from a $q^{-2}$ dependence are most easily detectable at large values of $q^2$. Technically these can be achieved by using colliding beams of electrons. Tests on the breakdown of quantum electrodynamics in this manner (report by Ting, 1968) using beams of 550 MeV electrons have shown excellent agreement between experiment and theory; the quantitative discrepancy was evaluated as

$$\frac{1}{\Lambda^2} = (-0{\cdot}06 \pm 0{\cdot}06)\ \text{GeV}^2.$$

Ignoring signs one could say $\Lambda \lesssim 4$ GeV, and if one Fourier transformed this value Coulomb's law is holding down to distances of $\sim 10^{-14}$ cm.

By far the most spectacular demonstration of the validity of quantum electrodynamics is the series of $g-2$ experiments performed at CERN.

These depend on the fact that the $g$ value‡ of the muon is slightly larger than two, consequently the magnetic moment (spin) of a moving muon rotates slightly more rapidly than the particle's orbital motion when placed in a magnetic field $B$ (denoted by the cross in Fig. 5.4)

$$\omega_c = \frac{e}{m} B \qquad \omega_L = g \frac{e}{2m} B \tag{5.2.1}$$

where $\omega_c$ and $\omega_L$ are the cyclotron and Larmor precession frequency respectively. The relevant diagrams for calculating $g$ are given in Fig. 5.4

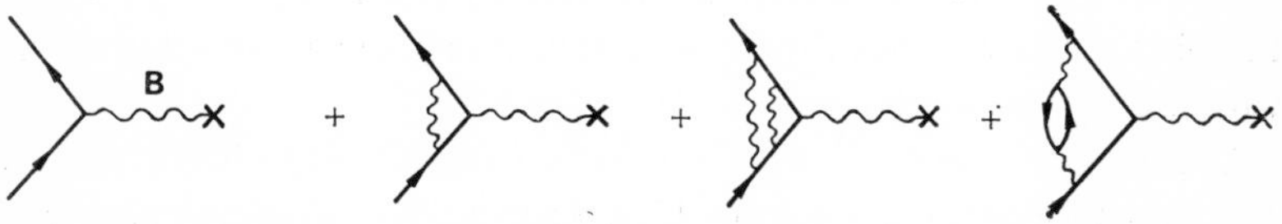


FIG. 5.4. Contributions to the magnetic moment of the muon (these occur for any Dirac particle).

where the first one on the left provides $g = 2$. The remaining diagrams give

$$\frac{g-2}{2} = \frac{\alpha}{2\pi} + 0{\cdot}7658 \frac{\alpha^2}{\pi^2} + 2{\cdot}49 \frac{\alpha^3}{\pi^3} + \cdots$$

$$= (11{,}655{\cdot}7) \times 10^{-7}$$

whilst experiment (Bailey *et al.*, 1968) gives

$$\frac{g-2}{2} = (11{,}661{\cdot}6 \pm 3{\cdot}1) \times 10^{-7}.$$

The deviation of $(5{\cdot}9 \pm 3{\cdot}1) \times 10^{-7} = (480 \pm 270)$ ppm $\equiv \Lambda \gtrsim 5$ GeV cannot be regarded as significant.

It is of interest to note that if the accuracy of the experiment could be increased by an order of magnitude, then strong interaction effects should be detectable. These would manifest themselves through a diagram like

‡ $g$ is defined as

$$\boldsymbol{\mu} = g \frac{e\hbar}{2mc} \mathbf{s}$$

where $\boldsymbol{\mu}$ and $\mathbf{s}$ are the magnetic moment and spin respectively.

that on the extreme right of Fig. 5.4 with a $\varrho$-meson replacing the electron pair (the $\varrho$ and $\gamma$ have the same quantum numbers). Attempts are under way to reach these limits.

## 5.3. Electromagnetic Form Factors

### 5.3(a). Nucleon form factors

In contrast with the apparently exact relationship between calculation and experiment in the interaction of leptons with electromagnetic radiation a totally different situation arises when hadrons are involved. In effect the lepton has a point-like structure and the photon interacts with the lepton at a point.

Consider the interaction of electrons with protons. If we used the appropriate Feynman diagram [Fig. 5.5(b)] and calculated the cross-

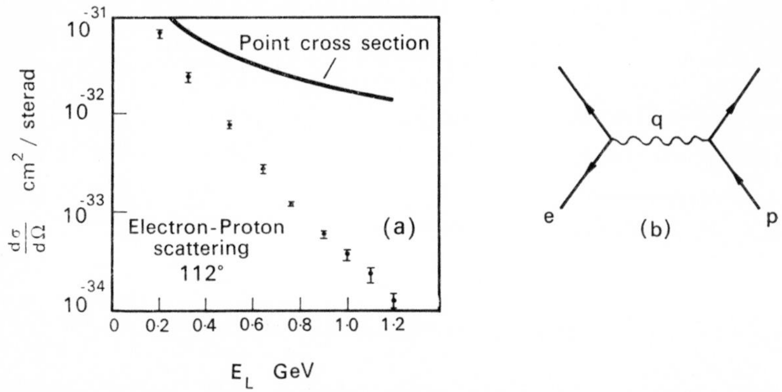


FIG. 5.5.(a) Cross-section for elastic *ep* scattering at 112° as a function of laboratory energy for the electron (Olson *et al.*, 1961). The experimental points are given together with the result expected from treating the proton as a point particle as in the diagram (b).

section for scattering at high energies then the deviations between calculation and experiment become enormous at high energies [Fig. 5.5(a)].

The discrepancy arises because at high energies the photon no longer "sees" the proton as a point particle but as a diffuse structure (Fig. 5.6).

This structure can be Fourier analysed into four momentum space and one can then write

$$T_{fi} = e^2 \bar{u}_{e'} \gamma_\mu u_e \frac{1}{q^2} \langle p' | j_\mu | p \rangle \tag{5.3.1}$$

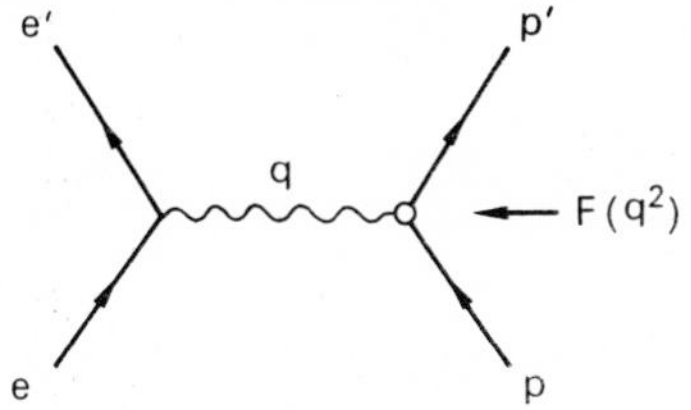


FIG. 5.6.

where the term $\langle p' | j_\mu | p \rangle$ is the most general four vector which can be constructed. This turns out to be‡

$$\langle p' | j_\mu | p \rangle = \bar{u}_{p'} \left[ \gamma_\mu F_1(q^2) + \frac{\sigma_{\mu\nu} q_\nu}{2m_p} F_2(q^2) \right] u_p. \tag{5.3.2}$$

All other four vectors can be reduced to this expression or are not gauge invariant. In the above expression $m_p$ in the mass of the proton and $F_1$ and $F_2$ are called form factors; physically the second term is identified with the anomalous part of the nucleon magnetic moment. If a proton had no strong interaction properties, then one might expect its magnetic moment to be 1 nuclear magneton

$$\mu_p = g \frac{e}{2m_p} s; \quad g = 2, \quad s = \tfrac{1}{2}$$

and calculable in the same way as for leptons. Instead it is 2·79 magnetons, so in the limit $q^2 \to 0$ one would expect

$$F_1(q^2) \to 1 \qquad F_2(q^2) \to 1 \cdot 79.$$
$$q^2 \to 0 \qquad q^2 \to 0.$$

Whilst $F_1$ and $F_2$ give the simplest way of reducing $j_\mu$, it turns out that their direct application is not the simplest way of analysing the experi-

‡ Details of the derivation may be found in § 11.4 of *P.E.P.*

mental data. Instead one uses the linear combinations

$$G_M(q^2) = F_1(q^2) + F_2(q^2) \to 2{\cdot}79 \quad \text{for} \quad q^2 \to 0$$

$$G_E(q^2) = F_1(q^2) - \frac{q^2}{4m_p^2} F_2(q^2) \to 1 \quad \text{for} \quad q^2 \to 0.$$

These are known as the magnetic and electric form factors, and the expression for the cross-section becomes

$$\frac{d\sigma}{d\Omega} = \sigma_M \left[ \frac{1}{1 + q^2/4m_p^2} \left( G_E^2 + \frac{q^2}{4m_p^2} G_M^2 \right) + \frac{2q^2}{4m_p^2} G_M^2 \tan^2 \frac{\vartheta}{2} \right] \tag{5.3.3}$$

where $\sigma_M$ is the cross-section obtained if the proton is treated as a point Dirac particle. The equation is known as the Rosenbluth formula.

The usual way of analysing data is to plot

$$\frac{1}{\sigma_M} \frac{d\sigma}{d\Omega} \cot^2 \frac{\vartheta}{2}$$

for fixed $q^2$ against $\cot^2 \vartheta/2$ and a straight line is expected and indeed found. Typical results are displayed in Fig. 5.7 (Albrecht *et al.*, 1967). $G_M^2$ can then be found from the intercept and $G_E^2$ from the slope.

Neutron form factors can also be measured since the neutron has an anomalous magnetic moment ($-1{\cdot}91$ magnetons) even if it is uncharged. The measurements are carried out in deuterium and proton data subtracted.

An enormous amount of work has been done on the nucleon form factors during the past ten years and the data seems capable of reduction to two very simple expressions (the units for $q^2$ are GeV$^2$; $\mu_p$ and $\mu_n$ are the magnetic moments of the proton and neutron respectively)

$$G_{E_p}(q^2) = \frac{G_{M_p}(q^2)}{\mu_p} = \frac{G_{M_n}(q^2)}{\mu_n} \sim \frac{1}{(1 + q^2/0{\cdot}71)^2} \tag{5.3.4}$$

$$G_{E_n}(q^2) \sim 0 \qquad q^2 = 0 \to 25 \text{ GeV}^2.$$

Some problems exist for the determination of $G_{E_n}(q^2)$ as $q^2 \to 0$, but the above expression is a reasonably good approximation.

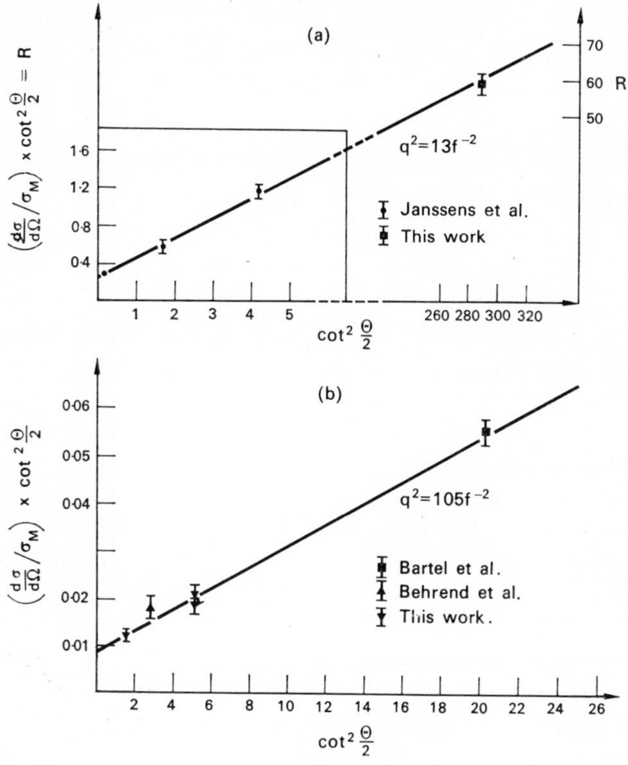


FIG. 5.7.

No satisfactory interpretation of the first expression ("the dipole formula") has been made. Since the vector mesons ($\varrho$, $\omega$, $\phi$) have spin-parity $1^-$ like the photon, a first approximation to accounting for the form factor might be to say that the photon couples to the vector mesons in the meson cloud about the nucleon (Fig. 5.8). This would give a term of the type

$$G \sim \frac{1}{1 + q^2/m_\varrho^2} \sim \frac{1}{1 + q^2/0{\cdot}57}.$$

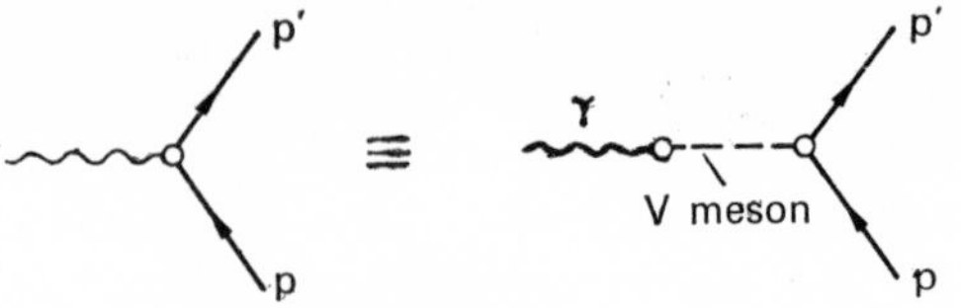


FIG. 5.8.

This is obviously wrong. One attractive speculation is that a further set of vector mesons exist with relative coupling $-\frac{1}{2}$

$$G \sim \frac{2}{1 + q^2/m_\varrho^2} - \frac{1}{1 + q^2/2m_\varrho^2} = \frac{1}{(1 + q^2/m_\varrho^2)\,(1 + q^2/2m_\varrho^2)}.$$

This would give about the right $q^2$ dependence, but no evidence exists for the heavier vector mesons.

### 5.3(b). Pion form factors

If $\varrho$-mesons are to contribute significantly to the electromagnetic form factors the most obvious place to look is in electron–pion scattering.

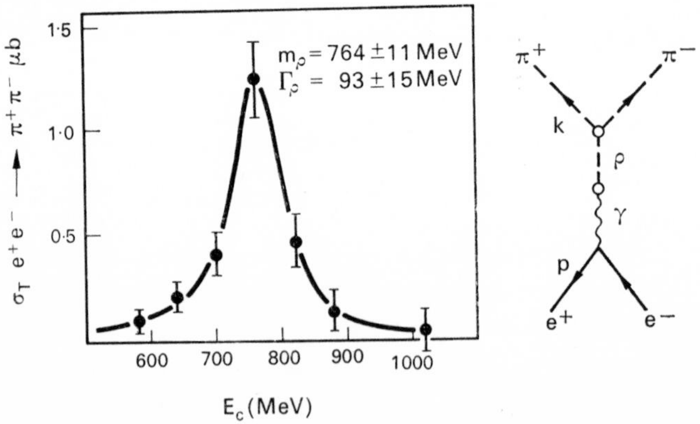


FIG. 5.9. Yield of $\pi^+\pi^-$ pairs from $e^+e^-$ annihilation as a function of energy in the c-system (Auslander *et al.*, 1967) with graph illustrating the dominant amplitude in the process.

Experiments with colliding beams $e^+e^- \to \pi^+\pi^-$ do indeed show $\varrho$ dominance. The measured cross-section peaks strongly at energies corresponding to the $\varrho$ mass (Fig. 5.9), and so the diagram on the right of the figure appears to be a justifiable assumption. Consider the cross-section for this process. Rather than evaluate the Dirac $\gamma$-matrices we shall use the principle of detailed balance (§ 2.4). If we denote the irrelevant kinematic factor $(2\pi E_c n)^{-2}$ by $C$ and call the momentum of the pions and electrons in the c-system $k$ and $p$ respectively, then we can write our cross-sections in the manner of equations (2.4.3) and (2.4.5)—we have dropped the summation signs for convenience of writing

$$\sigma(e^+e^- \to \pi^+\pi^-) = \frac{C}{(2s_e+1)^2}\frac{k}{p}\int d\Omega\, |T_{fi}|^2$$

$$\sigma(\pi^+\pi^- \to e^+e^-) = C\frac{p}{k}\int d\Omega\, |T_{if}|^2$$

$$= \sigma(\pi^+\pi^- \to \varrho \to \pi^+\pi^-)\frac{\Gamma_{ee}}{\Gamma_{\pi\pi}}$$

therefore

$$\sigma(e^+e^- \to \pi^+\pi^-) = \frac{1}{4}\frac{k^2}{p^2}\sigma(\pi\pi \to \varrho \to \pi\pi)\frac{\Gamma_{ee}}{\Gamma_{\pi\pi}} \tag{5.3.5}$$

where $\Gamma_{\pi\pi}$ and $\Gamma_{ee}$ are the decay rates of the $\varrho$-meson to pions and electrons respectively. The total decay rate $\Gamma$ is given by

$$\Gamma = \Gamma_{\pi\pi} + \Gamma_{ee} + \Gamma_{\mu\mu} \sim \Gamma_{\pi\pi}$$

since $\Gamma_{\pi\pi}$ is a strong interaction. The branching ratio of the $\varrho$ to electron pairs can therefore be written as

$$B = \frac{\Gamma_{ee}}{\Gamma_{\pi\pi}}. \tag{5.3.6}$$

All we have done in equation (5.3.5) is to say that if the $\varrho$-meson couples to both $\pi$ and $e$ pairs then we can replace the lower part of the diagram $ee \to \pi\pi$ in Fig. 5.9 by the upper part and multiply by the branching ratio. Now consider the process $\pi\pi \to \varrho \to \pi\pi$; we have previously shown

(4.2.1) that the matrix element for $\varrho$-decay (Fig. 5.10) is

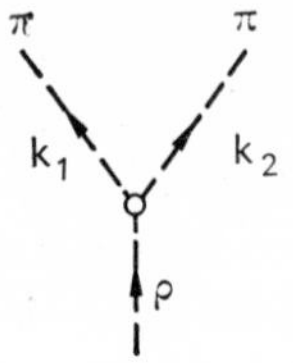


$$T_{fi} = \gamma\, e_\mu (k_1 - k_2)_\mu .$$

FIG. 5.10.

Hence for $\pi\pi \to \varrho \to \pi\pi$ we have [compare (4.3.11)]

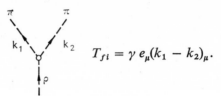

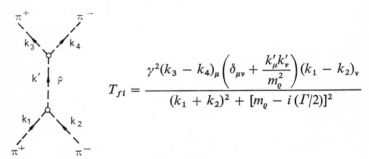

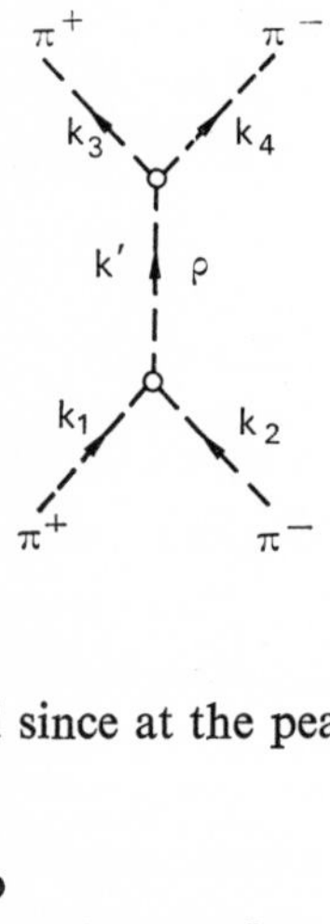


$$T_{fi} = \frac{\gamma^2 (k_3 - k_4)_\mu \left(\delta_{\mu\nu} + \dfrac{k'_\mu k'_\nu}{m_\varrho^2}\right)(k_1 - k_2)_\nu}{(k_1 + k_2)^2 + [m_\varrho - i\,(\Gamma/2)]^2}$$

FIG. 5.11.

and since at the peak of the cross-section in the $c$-system

$$k' = k_1 + k_2 = 0, \quad im_\varrho$$

also

$$\mathbf{k}_1 = -\mathbf{k}_2; \quad \mathbf{k}_3 = -\mathbf{k}_4; \quad |\mathbf{k}_1| = |\mathbf{k}_2| = |\mathbf{k}_3| = |\mathbf{k}_4| = k$$

then (neglecting $\Gamma^2/4$)

$$T_{fi} = \frac{\gamma^2 (\mathbf{k}_3 - \mathbf{k}_4)\cdot(\mathbf{k}_1 - \mathbf{k}_2)}{-m_\varrho^2 + (m_\varrho - i\Gamma/2)^2} = \frac{\gamma^2\, 4k^2 \cos\vartheta}{-i\Gamma m_\varrho} \tag{5.3.7}$$

Thus our cross-section becomes

$$\sigma(\pi\pi \to \varrho \to \pi\pi) = \frac{1}{(2\pi m_\varrho)^2}\,\frac{1}{16}\int d\Omega\, |T_{fi}|^2 = \frac{1}{3\pi}\,\frac{\gamma^4 k^4}{m_\varrho^4 \Gamma^2}$$

and by (5.3.5) and (5.3.6)

$$\sigma(ee \to \pi\pi) = \frac{1}{4}\frac{k^2}{p^2}\frac{1}{3\pi}\frac{\gamma^4 k^4}{m_\varrho^4 \Gamma^2} B = \frac{1}{3\pi}\frac{\gamma^4 k^6}{m_\varrho^6 \Gamma^2} B$$

since $2p = m_\varrho$ at the peak of the cross-section. Now $\gamma$ can be eliminated by calculating $\Gamma_{\pi\pi}$ [compare (2.3.8)] in the rest frame of the $\varrho$-meson

$$\Gamma_{\pi\pi} = (2\pi)^4 \frac{1}{8m_\varrho} \overline{\sum_i} \sum_f \int d\Omega\, |T_{fi}|^2 \frac{1}{(2\pi)^6} \frac{k}{m_\varrho}$$

$$= \frac{\gamma^2}{(2\pi)^2} \frac{1}{8m_\varrho} \overline{\sum_i} \sum_f \int d\Omega\, |e_\mu (k_1 - k_2)_\mu|^2 \frac{k}{m_\varrho}$$

$$= \frac{\gamma^2}{4\pi} \frac{2k^3}{3m_\varrho^2} \tag{5.3.8}$$

hence since $\Gamma \sim \Gamma_{\pi\pi}$

$$\sigma(ee \to \pi\pi) = \frac{1}{3\pi} \frac{\gamma^4 k^6}{m_\varrho^6} \frac{16\pi^2 9 m_\varrho^4}{\gamma^4 4k^6} B = \frac{12\pi}{m_\varrho^2} B. \tag{5.3.9}$$

Inserting the measured value for $B$ of $\sim 5 \times 10^{-5}$, a cross-section of $\sim 1\ \mu$b results; this agrees satisfactorily with the measured value. It should be noted that we have slightly oversimplified matters here, since the other vector mesons can also couple to the photon (compare Fig. 5.8) and hence contribute to the intermediate state. Of these the $\omega$ contribution is the most important since its mass almost coincides with that of the $\varrho$ meson. Evidence of $\omega - \varrho$ interference in $e^+e^- \to \pi^+\pi^-$ has been obtained (Augustin *et al.*, 1969a‡); as the $\omega$ amplitude is relatively weak the overall picture presented above is virtually unchanged.

If the intermediate $\varrho$ state is taken seriously then we have a model for the electric form factor of the pion. Consider elastic $e\pi$ scattering; the appropriate diagram is shown in Fig. 5.12 and the associated amplitude

‡ The experiments of this group have also demonstrated the occurrence of the processes $e^+e^- \to \omega \to \pi^+\pi^-\pi^0$ and $e^+e^- \to \phi \to K\bar{K}$.

is‡

$$T_{fi} = e^2 \bar{u}_{e'} \gamma_\mu u_e \frac{1}{q^2} F_\pi(q^2) (k_1 + k_2)_\mu \tag{5.3.10}$$

where $q = k_2 - k_1$; for elastic scattering $q^2 > 0$. The form factor from Fig. 5.12 is

$$F_\pi(q^2) = \frac{g_{\varrho\gamma}\gamma}{q^2 + (m_\varrho - i\Gamma/2)^2} \tag{5.3.11}$$

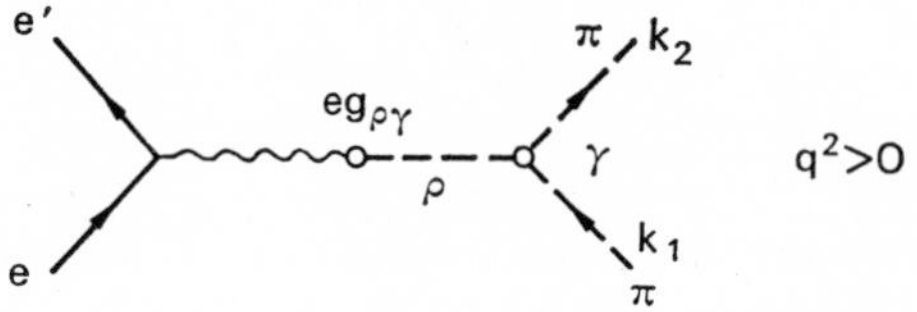


FIG. 5.12.

where $g_{\varrho\gamma}$ contains all the numerator terms apart from $\gamma$. Now in the limit $q^2 \to 0$ we require $F_\pi(q^2) \to 1$; thus if we ignore the small imaginary term at $q^2 = 0$

$$F_\pi(q^2) = \frac{m_\varrho^2}{q^2 + (m_\varrho - i\Gamma/2)^2}. \tag{5.3.12}$$

This result would imply that $F_\pi(q^2)$ has the form shown in Fig. 5.13 (note the difference in scales).

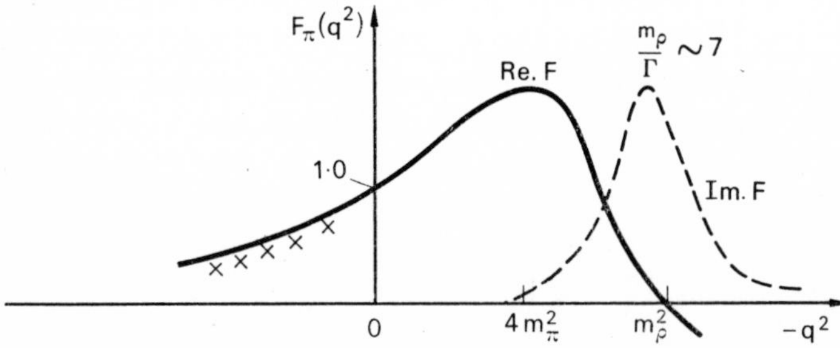


FIG. 5.13.

‡ The choice of $(k_1 + k_2)_\mu$ or $(k_1 - k_2)_\mu$ in (5.3.5) is determined by conservation of four momentum, $p + k_1 = p' + k_2$, plus the Dirac equation—compare (4.3.14).

The region for $q^2 < -4m_\pi^2$ has been explored in the colliding $e^+e^-$ beam experiments discussed above. The amplitude for $e^+e^- \to \pi^+\pi^-$ can be written as

$$T_{fi} = -e^2\, \bar{v}\gamma_\mu u \frac{1}{q^2} F_\pi(q^2)\,(k_1 - k_2)_\mu \tag{5.3.13}$$

and evaluation of the Dirac $\gamma$-matrices leads to the cross-section

$$\sigma(e^+e^- \to \pi^+\pi^-) = \frac{2}{3}\pi\alpha^2 \frac{k_\pi^3}{E_c^5} |F_\pi|^2 \tag{5.3.14}$$

where $k_\pi$ is the momentum of the pion in the $c$-system.

Experimentally $F_\pi(q^2)$ for $q^2 > 0$ has been determined from inelastic electron scattering under suitable kinematic conditions (Fig. 5.14). The agreement between experiment and the $\varrho$ model is reasonable, but the experimental results appear to be systematically low as indicated in Fig. 5.13 by the crosses (see Panofsky (1968) for a discussion of experimental data).

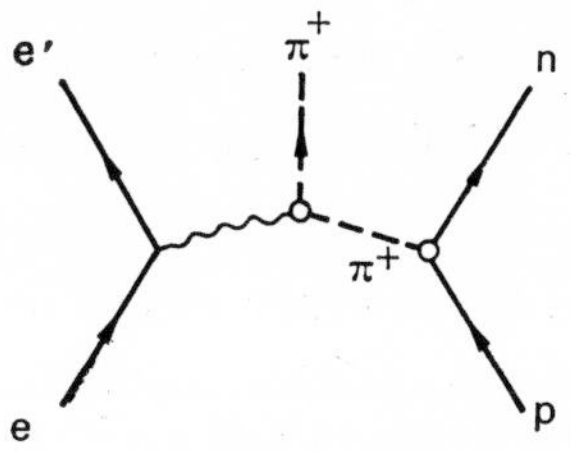


FIG. 5.14.

CHAPTER 6

# WEAK INTERACTIONS

APART from their strength the most notable feature of weak interactions is their behaviour under the operations of spatial reflection (parity) and charge conjugation. We shall examine these transformations first of all.

## 6.1. The Operations $P$, $C$, $T$

The transformations applied to quantum mechanical systems are of two main types:

1. additive or continuous in which a finite transformation can be built by adding a series of infinitesimal transformations—rotation is an example of this;
2. multiplicative or discrete—these cannot be built up from infinitesimal transformations. The most important of these, the operations of spatial reflection ($P$), charge conjugation ($C$) and time reserval ($T$), return the system to its original state upon double application

$$P^2 = C^2 = T^2 = \hat{1}. \tag{6.1.1}$$

### 6.1(a). Parity

The operation of spatial reflection changes $\mathbf{x}$ to $-\mathbf{x}$, which in turn causes momenta to change from $\mathbf{p}$ to $-\mathbf{p}$. Angular momentum $(\mathbf{r} \times \mathbf{p})$ is therefore left unchanged by the operation. We can therefore represent the action of the operation on a particle of momentum $\mathbf{k}$ and spin $\mathbf{s}$ by

$$P\,|\mathbf{k}, \mathbf{s}\rangle = \zeta_P\,|-\mathbf{k}, \mathbf{s}\rangle \tag{6.1.2}$$

and since a further application returns a particle to its original state

$$P^2 |\mathbf{k}, \mathbf{s}\rangle = \zeta_P P |-\mathbf{k}, \mathbf{s}\rangle = \zeta_P^2 |\mathbf{k}, \mathbf{s}\rangle = |\mathbf{k}, \mathbf{s}\rangle$$

then $\zeta_P^2 = 1$, and systems with $\zeta_P = +1$ or $-1$ are said to have even or odd parity respectively. For a single particle at rest $\zeta_P$ is called the intrinsic parity of the particle.

In the case of bosons intrinsic parity can be defined unambiguously by conservation arguments. Since fermions have half integral spin, conservation of angular momentum forces them to occur in pairs, and one is forced to define the relative parity of pairs of fermions.

Consider how this works for two Dirac particles; previously we examined Lorentz transformations for the Dirac equation [§ 3.5(d)], and decided that for a transformation of the type

$$x'_\lambda = a_{\lambda\nu} x_\nu \qquad \psi'(x') = U\psi(x)$$

then

$$U^{-1}\gamma_\lambda U = a_{\lambda\nu}\gamma_\nu. \tag{6.1.3}$$

A reflection of spatial coordinates means

$$\begin{aligned} a_{\lambda\nu} &= -1 \qquad \lambda = \nu = 1, 2, 3 \\ &\phantom{=}\; +1 \qquad \lambda = \nu = 4 \\ &\phantom{=}\;\;\; 0 \qquad \lambda \neq \nu. \end{aligned}$$

Thus the solution of (6.1.3) for a spatial reflection is

$$U = \zeta_P \gamma_4$$

where $\zeta_P$ is a phase factor ($e^{i\phi}$), and so

$$\psi'(-\mathbf{x}, t) = \zeta_P \gamma_4 \psi(\mathbf{x}, t).$$

Consider the implications of this result for spinors [§ 3.5(e)]. We can write in $2 \times 2$ shorthand [compare (3.5.35) and (5.1.2)]

$$u = \sqrt{\frac{m+E}{2m}} \begin{pmatrix} 1 \\ \dfrac{\boldsymbol{\sigma} \cdot \mathbf{p}}{E+m} \end{pmatrix} \qquad v = \sqrt{\frac{m+E}{2m}} \begin{pmatrix} \dfrac{\boldsymbol{\sigma} \cdot \mathbf{p}}{E+m} \\ 1 \end{pmatrix} \tag{6.1.4}$$

hence

$$\begin{aligned} u_{\mathbf{p}} &\xrightarrow{P} u_{-\mathbf{p}} = \gamma_4 u_{\mathbf{p}} \\ v_{\mathbf{p}} &\quad\; v_{-\mathbf{p}} = -\gamma_4 v_{\mathbf{p}} \end{aligned} \qquad \text{since} \quad \gamma_4 = \begin{pmatrix} 1 & 0 \\ 0 & -1 \end{pmatrix}.$$

Since $u$ and $v$ are associated with particle and antiparticle respectively, then they must have opposite intrinsic parity.

Now consider systems of two particles moving in their mutual $c$-system with momentum $k$ and orbital angular momentum $l$. Then the completeness relation (Appendix A3) gives us

$$|klm\rangle = \sum_{\vartheta,\phi} |k,\vartheta,\phi\rangle\,\langle k,\vartheta,\phi \mid klm\rangle = \sum_{\vartheta,\phi} Y_l^{*m}(\vartheta,\phi)\,|\mathbf{k};\,-\mathbf{k}\rangle$$

and

$$\begin{aligned} P\,|klm\rangle &= \zeta_1\zeta_2 \sum_{\vartheta,\phi} Y_l^{*m}(\pi-\vartheta,\phi+\pi)\,|-\mathbf{k};\,\mathbf{k}\rangle \\ &= \zeta_1\zeta_2 \sum_{\vartheta,\phi} (-1)^l\, Y_l^{*m}(\vartheta,\phi)\,|\mathbf{k};\,-\mathbf{k}\rangle \qquad \text{since we sum over all directions} \\ &= \zeta_1\zeta_2(-1)^l\,|klm\rangle \end{aligned} \tag{6.1.5}$$

This rule implies that a system decaying to two pions must obey the so-called natural parity sequence ($\zeta_1\,\zeta_2 = 1$ for two pions)‡

$$j^P = 0^+, 1^-, 2^+, 3^- \cdots \qquad \pi^+\pi^- \quad \text{or} \quad \pi^\pm\pi^0. \tag{6.1.6}$$

If both pions have the same charge then a further restriction occurs. The interchange of a pair of pions is equivalent to spatial reflection in the $c$-system. The requirements of Bose–Einstein statistics then force the new wave function to be even for a pair of identical pions, thus pairs of identical pions must belong to the configurations

$$j^P = 0^+, 2^+, 4^+ \cdots \qquad \pi^+\pi^+, \pi^-\pi^-, \pi^0\pi^0. \tag{6.1.7}$$

‡ The intrinsic parity of the pion is negative. This conclusion follows from the observation of the reaction $\pi^- d \to nn$ for pions at rest ($s$-state capture). Since the deuteron has spin-parity $1^+$ then $j^P$ for the initial state is $1^{\zeta\pi}$. The requirements of Fermi–Dirac statistics imply that the only possible state for two neutrons with $j = 1$ is ${}^{2s+1}L_1^{\zeta P} = {}^3P_1^-$, hence $\zeta\pi = -1$.

### 6.1(b). Charge conjugation

The operation of charge conjugation changes a particle (charge $Q$) to an antiparticle ($-Q$) whilst leaving its mechanical properties unchanged

$$C\,|\mathbf{k}, \mathbf{s}, Q\rangle = \zeta_C\,|\mathbf{k}, \mathbf{s}, -Q\rangle.$$

This operation carries with it the implication that magnetic moments change sign and that the Gell-Mann, Nishijima relation (1.4.3)

$$\frac{Q}{e} = I_3 + \frac{1}{2}(B + S)$$

transforms as $I_3 \to -I_3$, $B \to -B$ and $S \to -S$.

The operations of charge conjugation and spatial exchange are physically equivalent for a pair of spinless particles with opposite charge in their $c$-system; from (6.1.5) we therefore have the relation

$$C\,|klm\rangle = \zeta_1\zeta_2\,(-1)^l\,|klm\rangle \tag{6.1.8}$$

and in general $\zeta_1\zeta_2 = |\zeta_C|^2 = 1$. If the particles have spin, complications occur. For two identical fermions the requirements of Fermi–Dirac statistics lead to the antisymmetry relation

$$X = X_r X_s = (-1)^l\,(-1)^{s+1} = -1 \tag{6.1.9}$$

where $X_r$ refers to spatial exchange and $X_s$ to spin exchange. Now consider the following sequence of operations for a proton–antiproton pair

$$\begin{array}{lccccccc} \text{spins} & \uparrow\ \downarrow & \xrightarrow{X_r} & \downarrow\ \uparrow & \xrightarrow{C} & \downarrow\ \uparrow & \xrightarrow{X_s} & \uparrow\ \downarrow \\ & p\ \bar{p} & & \bar{p}\ p & & p\ \bar{p} & & p\ \bar{p}. \end{array}$$

Thus we have a "new" system indistinguishable from the original one and so

$$X_r C X_s = -1 \qquad C = (-1)^{l+s}. \tag{6.1.10}$$

Two bosons can be treated in a similar way. Here the spin exchange property is $X_s = (-1)^s$ and so

$$X_r C X_s = +1 \qquad C = (-1)^{l+s}. \tag{6.1.11}$$

Certain particles, e.g. $\gamma$, $\pi^0$, are self-conjugate under the charge conjugation operation. Nevertheless they can have an intrinsic charge parity

$$C\,|\gamma\rangle = \zeta_C\,|\gamma\rangle \qquad C\,|\pi^0\rangle = \zeta_C\,|\pi^0\rangle.$$

If the particles are considered in isolation nothing can be said about the value of $\zeta_C$. However, since the photons couple to electric currents (4.3.12), and the charge conjugation operation by definition causes the change $j_\mu \to -j_\mu$, then $\zeta_C$ must be $-1$ for the photon to preserve invariance of the electromagnetic interaction under charge conjugation. Consideration of the strong interaction of the $\pi^0$ meson with baryons leads to the requirement $\zeta_C = +1$ by similar arguments (the same conclusion can also be reached from the observed decay mode $\pi^0 \to 2\gamma$ since $\zeta_C = +1$ for $2\gamma$). Thus we can write

$$C\,|\gamma\rangle = -|\gamma\rangle \qquad C\,|\pi^0\rangle = |\pi^0\rangle. \tag{6.1.12}$$

### 6.1(c). Time reversal

The last discrete transformation is time reversal. Under this operation momentum reverses, $\mathbf{p} \to -\mathbf{p}$, and so also does angular momentum, $(\equiv \mathbf{r} \times \mathbf{p})$. We also want initial states to become final states and vice versa, hence

$$T\,|\mathbf{k}, \mathbf{s}\rangle = \zeta_T\,\langle -\mathbf{s}, -\mathbf{k}|. \tag{6.1.13}$$

## 6.2. Invariance Properties in Weak Interactions

### 6.2(a). Introduction

Let us assume that we can split the $S$ operator [§ 2.2(a)] into two parts

$$S = S_1 + S_2 \tag{6.2.1}$$

such that under a discrete transformation $U(\equiv P, C, T)$

$$U^{-1}SU = S_1 - S_2 \neq S. \tag{6.2.2}$$

Thus $S_1$ and $S_2$ represent invariant and non-invariant parts of the $S$ operator respectively.

The realisation that $S_2 \neq 0$ in weak interactions came from a study of $K$ decay. Early measurements established the equality of mass and lifetime for two different decay modes

$$\underset{\vartheta \text{ decay}}{K^+ \to \pi^+\pi^0} \qquad \underset{\tau \text{ decay}}{K^+ \to \pi^+\pi^+\pi^-}$$

Now for $K \to 2\pi$, $j^P$ for the two-pion system is $0^+$, $1^-$, $2^+$ ... [compare (6.1.6)], and for $K \to 3\pi$ we can represent the pions in the following way:

$$\underset{j}{K^+} \to \underbrace{\pi^-\overbrace{(\pi^+\pi^+)}^{L}}_{l} \qquad \mathbf{j} = \mathbf{L} + \mathbf{l}.$$

The two $\pi^+$ system can only adopt the configurations $0^+, 2^+, 4^+$ ... because of Bose–Einstein statistics (6.1.7); a single pion has odd intrinsic parity and so the three-pion configuration has possible spin-parity assignments

$$\begin{array}{lll} j^P = 0^-, & 1^+, & 1^- \ldots \\ l = 0 \;\; L = 0 & l = 1 \;\; L = 0 & l = 2 \;\; L = 2 \end{array} \tag{6.2.3}$$

Now the spin of the kaon is zero [§ 7.1 (c)], and so one had the apparently contradictory situation that the kaon could decay into systems of odd and even parity, implying that its quantum numbers were simultaneously $0^-$ and $0^+$ if parity was conserved.

This dilemma was resolved by Lee and Yang (1956a) who pointed out that whilst there was good evidence that parity was conserved in strong and electromagnetic interactions no equivalent evidence existed for weak interactions.

Now consider the problem of detecting $S_2$ in our $S$ operator (6.2.1)

$$S = S_1 + S_2.$$

One way is to work in a situation where $S_1 = 0$, for example upper limits have been placed on parity violation in strong interactions by looking for nuclear transitions which can only take place if parity conservation is violated.

Such a situation is displayed in Fig. 6.1; the 8·88 MeV level of $O^{16}$ has a spin-parity $2^-$. Its decay via $\alpha$-particle emission to the ground state of $C^{12}$ was sought (Boyd *et al.*, 1968). Since the $\alpha$-particle has spin-

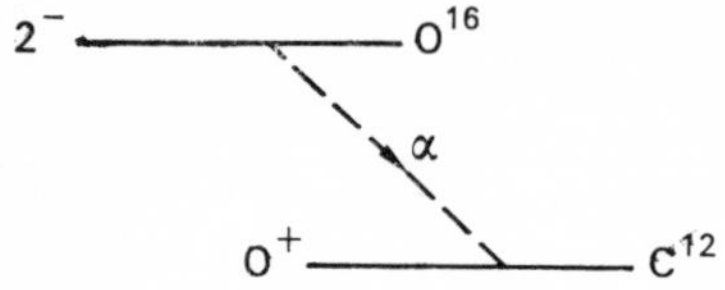


$l = 2$ for $\alpha$ conserves $j$, breaks $P$.

FIG. 6.1.

parity $0^+$ conservation of angular momentum in this reaction can only be achieved at the expense of a failure of parity conservation. The experiment led to the limit

transition rate $|S_2|^2 \leqq 3 \times 10^{-13}$ (rate expected for a strong interaction).

Another way to detect $S_2$ is to measure the interference term between $S_1$ and $S_2$

$$|S|^2 = |S_1|^2 + |S_2|^2 + 2\mathrm{Re}S_1^*S_2. \tag{6.2.4}$$

Consider a situation where we choose an $S$ operator which leads to a matrix element

$$T_{fi} = a + b\,\mathbf{S}\cdot\hat{\mathbf{p}}$$

where $a$, $b$ are numerical coefficients, $\mathbf{S}$ a spin direction and $\hat{\mathbf{p}}$ a unit vector in the direction of a momentum. Now under a spatial reflection $\mathbf{S} \to \mathbf{S}$, $\hat{\mathbf{p}} \to -\hat{\mathbf{p}}$, hence

$$T_{fi} = a + b\,\mathbf{S}\cdot\hat{\mathbf{p}} \xrightarrow{P} a - b\,\mathbf{S}\cdot\hat{\mathbf{p}} \tag{6.2.5}$$

and $b = 0$ if parity is conserved, if not

$$|T_{fi}|^2 = |a|^2 + |b|^2 + 2\mathrm{Re}\,a^*b\,\mathbf{S}\cdot\hat{\mathbf{p}}.$$

Thus a failure of parity conservation ($b \neq 0$) would manifest itself as an anisotropic angular distribution of a particle about a spin direction. This was the principle used in the first experiments to search for the failure of parity conservation in weak interactions. The experiments involved nuclear $\beta$-decay and muon decay. In the former (Wu *et al.*, 1957) nuclei of $Co^{60}$ were polarised and the angular distribution about the direction of polarisation of the $\beta$-decay electrons was examined. Strong asymmetries were found implying a maximum violation of parity

$$a^2 \sim b^2.$$

A similar situation was found in $\mu$-decay (Friedman and Telegdi, 1957; Garwin *et al.*, 1957).

### 6.2(b). Nuclear β-decay

Let us consider how parity violating terms can arise in nuclear $\beta$-decay theory and also their implications. The decay process $N \to N'e^-\bar{\nu}$, where $N$ and $N'$ are nuclei, can be conveniently treated as $\nu N \to N'e^-$, since the emission of antiparticles and the absorption of particles are equivalent processes [§ 3.5(e)]. Our basic diagram is shown in Fig. 6.2; the associated amplitude is

$$T_{fi} = \sum_i C_i \langle N_i \rangle \bar{u}_e O_i u_\nu \tag{6.2.6}$$

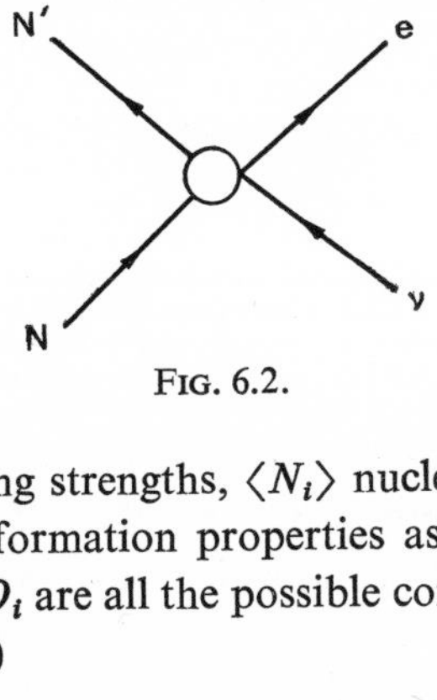


FIG. 6.2.

where the $C_i$ are coupling strengths, $\langle N_i \rangle$ nuclear matrix elements with the same Lorentz transformation properties as $O_i$ so that $T_{fi}$ remains Lorentz invariant, and $O_i$ are all the possible combinations of $\gamma$-matrices listed in equation (3.5.7)

$$O_i = \underset{\text{S}}{\hat{1}}, \underset{\text{V}}{\gamma_\mu}, \underset{\text{T}}{i\gamma_\mu\gamma_\nu}, \underset{\text{A}}{i\gamma_\mu\gamma_5}, \underset{\text{P}}{\gamma_5} \qquad \mu < \nu.$$

Now when we considered the Dirac equation for neutrinos [§ 3.5(f)] we found that

$$\mathbf{S} \cdot \mathbf{p}\, u_\nu = -\tfrac{1}{2}\gamma_5 E u_\nu .$$

It is apparent from this equation that if spatial coordinates are reflected $\mathbf{S} \cdot \mathbf{p}$ will change sign and so somehow $\gamma_5$ must be associated with a change of sign (more formally it is $\bar{\psi}\gamma_5\psi$ which has this property—see *P.E.P.* p. 85). Thus we may introduce a parity violating term in the matrix element for $\beta$-decay by inserting $\gamma_5$

$$T_{fi} = \sum_i C_i \langle N_i \rangle \bar{u}_e O_i u_\nu + \sum_i C_i' \langle N_i \rangle \bar{u}_e O_i \gamma_5 u_\nu$$

$$= \sum_i \frac{C_i + C_i'}{2} \langle N_i \rangle \bar{u}_e O_i (1 + \gamma_5) u_\nu$$

$$+ \sum_i \frac{C_i - C_i'}{2} \langle N_i \rangle \bar{u}_e O_i (1 - \gamma_5) u_\nu. \qquad (6.2.7)$$

Now the experimental observation of maximum parity violation implies that $|C_i|^2 = |C_i'|^2$, but we can go further. Since the neutrino is massless $(1 \pm \gamma_5) u_\nu$ implies neutrinos with spin parallel $(-)$ or antiparallel $(+)$ to their momentum respectively [§ 3.5(f)]. An ingenious experiment by Goldhaber, Grodzins and Sunyar (1958) showed that the neutrino spin and momentum were antiparallel. The experiment depended upon a measurement of the helicity of the photon emitted following $K$ electron capture in $Eu^{152}$

$$e^- + \mathrm{Eu}^{152} \rightarrow \mathrm{Sm}^{152*} + \nu$$
$$\downarrow$$
$$\mathrm{Sm}^{152} + \gamma.$$

The photons were detected by resonant scattering in $Sm^{152}$

$$\gamma + \mathrm{Sm}^{152} \rightarrow \mathrm{Sm}^{152*} \rightarrow \mathrm{Sm}^{152} + \gamma.$$

The kinematic conditions for resonant scattering are best fulfilled for those photons which emerge whilst the $Sm^{152*}$ nucleus is still recoiling from the neutrino emission, and which travel parallel to the direction of recoil. The helicity of the photons was determined by their ease of transmission through magnetised iron as a function of the field direction (Fig. 6.3). The helicity of the photon was found to be negative; since $Eu^{152}$ and $Sm^{152}$ both have spin zero, the helicity of the neutrino must also be negative in order to achieve overall conservation of linear and angular momentum. The data was compatible with 100 per cent polarisation of the neutrinos.

The requirement of neutrinos with negative helicity implies that we use $(1 + \gamma_5)u_\nu$, in equation (6.2.7). This result can be achieved by writing

$$C_i = C_i' = \frac{G_i}{\sqrt{2}} \tag{6.2.8}$$

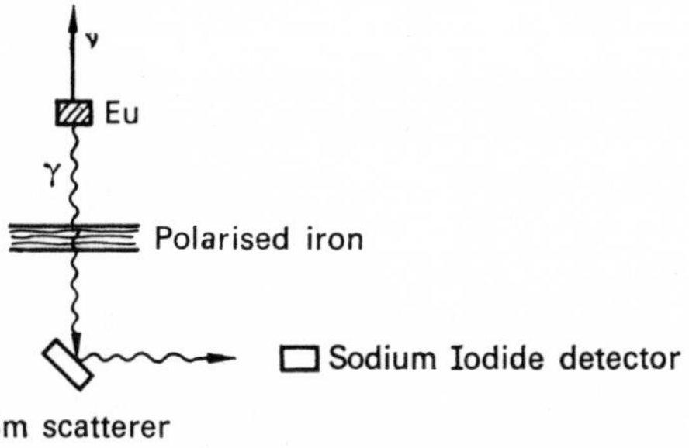


FIG. 6.3. Schematic outline of the experiment of Goldhaber *et al.* (1958) to determine the helicity of the neutrino.

where the $\sqrt{2}$ is a historical convention with no great significance. Thus our matrix element, (6.2.7), becomes

$$T_{fi} = \sum_i \frac{G_i}{\sqrt{2}} \langle N_i \rangle \, \bar{u}_e O_i (1 + \gamma_5) \, u_\nu . \tag{6.2.9}$$

Now we can write the $\gamma$-matrices in the above equation as

$$O_i(1 + \gamma_5) = \tfrac{1}{2}(1 - \gamma_5) \, O_i(1 + \gamma_5) + \tfrac{1}{2}(1 + \gamma_5) \, O_i(1 + \gamma_5)$$

and

$$\begin{aligned} \tfrac{1}{2}(1 - \gamma_5) \, O_i(1 + \gamma_5) &= 0 & i &= \text{S, T, P} \\ &= O_i(1 + \gamma_5) & i &= \text{V, A} \\ \tfrac{1}{2}(1 + \gamma_5) \, O_i(1 + \gamma_5) &= O_i(1 + \gamma_5) & i &= \text{S, T, P} \\ &= 0 & i &= \text{V, A} \end{aligned} \tag{6.2.10}$$

in addition

$$\bar{u}_e(1 \pm \gamma_5) = u_e^\dagger(1 \mp \gamma_5) \, \gamma_4 = [\gamma_4(1 \mp \gamma_5) \, u_e]^\dagger . \tag{6.2.11}$$

Thus, recalling the behaviour of the helicity states in § 3.5(f), we would expect

$i$ = V, A electron spin and momentum antiparallel if $m_e/E_e \to 0$

$i$ = S, T, P parallel.

Detailed calculation gives less than 100 per cent polarisation if the electron is non-relativistic; in this situation the polarisation $P$ reduces to $\pm v$ where $v$ is the electron velocity in units of $c$. Experiment‡ gave the result electron polarisation $P \sim -v$ thus showing that $\beta$-decay involves V, A couplings.

The observation of negative helicity for the electron also offers a qualitative explanation for the asymmetry observed in the $\beta$-decay of $Co^{60}$ [§ 6.2(a)]. The spins of the electron and antineutrino (positive helicity since $\nu$ has negative helicity) must balance the difference in spins of the initial and final states in $Co^{60}$. These are indicated by the broad arrows in Fig. 6.4; the helicities of the emitted particles then cause the

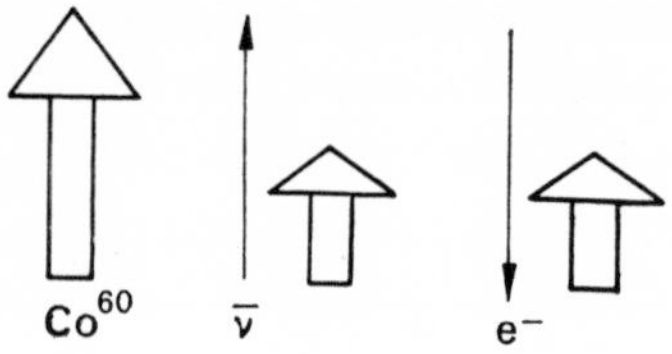


FIG. 6.4.

electrons to move in the direction indicated. Thus if we indicate the $Co^{60}$ polarisation by $\langle \mathbf{S}_N \rangle$ the matrix element must look roughly like

$$T_{fi} \sim 1 - \langle \mathbf{S}_N \rangle \cdot \hat{\mathbf{p}}$$

[compare (6.2.5)].

When experiments were performed with positron emission it was found that the positron's spin and momentum were parallel (positive helicity) and $P \sim +v$. Thus charge conjugation invariance was violated in $\beta$-decay. The experimental result can be understood within the context that positron

‡ A review of the experiments on electron polarisation may be found in Schopper (1966) and Wu and Moszkowski (1966).

emission involves the lepton pair $e^+\nu$ whereas electron emission involves $e^-\bar{\nu}$ [this statement is concerned with lepton conservation which we shall discuss in § 6.3(a)]. The decay process $N \to N'e^+\nu$ can be conveniently treated as $e^-N \to N\nu$ and since we require neutrinos with negative helicity a suitable matrix element is

$$T_{fi} = \sum_i \frac{G_i}{\sqrt{2}} \langle N_i \rangle \, \bar{u}_\nu (1 - \gamma_5) \, O_i u_e \qquad (6.2.12)$$

[compare (6.2.9) and (6.2.11)]. The observation of positive helicity for positrons again leads to the requirement of a V, A interaction by the argument developed in equations (6.2.10).

Finally consider nuclear part $\langle N_i \rangle$ of the $\beta$-decay matrix. If $A$ and $Z$ are atomic weight and number respectively, it will look like

$$\langle N_i \rangle = \langle N^A_{Z+1} | \sum_{j=1}^{A} \tau_j^+ O_{ij} \, | N^A_Z \rangle$$

where $\tau_j^+$ is the operator which transforms the nucleon $j$ from a neutron to a proton. Since neutrons and protons are Dirac particles moving at non-relativistic velocities we can write [compare (5.1.2)]

$$u = \begin{pmatrix} 1 \\ 0 \end{pmatrix} \qquad 1 = \begin{pmatrix} 1 & 0 \\ 0 & 1 \end{pmatrix} \qquad 0 = \begin{pmatrix} 0 & 0 \\ 0 & 0 \end{pmatrix}$$

hence from (3.5.4)

$$i = \mathrm{V} \quad u_p \gamma_\mu u_n \equiv (1 \;\; 0) \begin{pmatrix} 0 & -i\boldsymbol{\sigma} \\ i\boldsymbol{\sigma} & 0 \end{pmatrix} \begin{pmatrix} 1 \\ 0 \end{pmatrix}, \quad (1 \;\; 0) \begin{pmatrix} 1 & 0 \\ 0 & -1 \end{pmatrix} \begin{pmatrix} 1 \\ 0 \end{pmatrix} = 0, 1$$

$$i = \mathrm{A} \quad \bar{u}_p i \gamma_\mu \gamma_5 u_n \equiv (1 \;\; 0) \begin{pmatrix} 0 & \boldsymbol{\sigma} \\ -\boldsymbol{\sigma} & 0 \end{pmatrix} \begin{pmatrix} 0 & -1 \\ -1 & 0 \end{pmatrix} \begin{pmatrix} 1 \\ 0 \end{pmatrix},$$

$$(1 \;\; 0) \begin{pmatrix} i & 0 \\ 0 & -i \end{pmatrix} \begin{pmatrix} 0 & -1 \\ -1 & 0 \end{pmatrix} \begin{pmatrix} 1 \\ 0 \end{pmatrix} = -\boldsymbol{\sigma}, 0.$$

Thus the V term will not cause spin flip in the nucleus whilst the A term will. The components of the nuclear matrix element thus become

$$i = \mathrm{V} \quad \langle N_\mathrm{V} \rangle = \left\langle N^A_{Z+1} | \sum_j \tau_j \, | N^A_Z \right\rangle$$

$$i = \mathrm{A} \quad \langle N_\mathrm{A} \rangle = \left\langle N^A_{Z+1} | \sum_j \tau_j \sigma_j \, | N^A_Z \right\rangle$$

and are called Fermi and Gamow–Teller transitions respectively. Detailed studies of $e\nu$ correlations in nuclei containing varying mixtures of Fermi and Gamow–Teller strengths have confirmed the V, A requirement for $\beta$ decay.

### 6.2(c). μ- and π-decay

Studies of the electrons from $\mu$-decay have shown the same features as in nuclear $\beta$-decay—maximum violation of parity and charge conjugation invariance

$$\mu^+ \to e^+_+ \, \nu\bar{\nu}$$

$$\mu^- \to e^-_- \, \nu\bar{\nu}$$

where the upper indices refer to charge and lower to helicity. All the observed data on muon decay is consistent with a matrix element ($G_\mu$ is the coupling strength for muon decay)

$$T_{fi} = \frac{G_\mu}{\sqrt{2}} \bar{u}_\nu \gamma_\lambda(1 + \gamma_5) \, u_\mu \, \bar{u}_e \gamma_\lambda(1 + \gamma_5) \, u_\nu \qquad \mu^- \to e^- \, \nu\bar{\nu} \qquad (6.2.13)$$

that is a matrix element containing only vector (V) and axial vector (A) terms.‡ This process and its associated matrix element is of particular importance in the study of weak interactions since none of the particles is strongly interacting.

Now consider $\pi$-decay. We have shown previously (§ 2.5) that phase space arguments would lead us to expect a branching ratio or $\sim 3$ for $\pi \to e/\pi \to \mu$. In fact experiment shows it to be $10^{-4}$.

We write the interaction as

$$\pi^- \to l^- \bar{\nu} \qquad l \equiv e, \mu$$

and start from

$$T_{fi} = \sum_i \langle 0| \, T_i \, |\pi\rangle \, \bar{u}_l O_i (1 + \gamma_5) \, u_\nu .$$

‡ In fact the matrix element also implies that the relative signs of the V and A couplings are negative—this result may be easily obtained by starting from (6.2.9) with $i =$ V, A and an appropriate substitution for $\langle N_i \rangle$.

Allowing for the pseudoscalar nature of the pion only two terms are possible for $\langle 0 | T_i | \pi \rangle$ (the $\langle 0 |$ implies no final hadronic state)

$$\langle 0 | T_i | \pi \rangle = G_A \frac{p_\lambda}{m_\pi} \qquad i = A$$
$$= G_P \qquad i = P$$

where $p_\lambda$ is the four momentum of the pion and $m_\pi$ ensures that both $G_A$ and $G_P$ have the same dimensions. Then using the Dirac equations (3.5.29) and (3.5.11)

$$(i\gamma p + m)\, u = 0$$
$$\bar{u}(i\gamma p + m) = 0$$

we find for $m_\nu = 0$

$$T_{fi} = G_A \frac{p_\lambda}{m_\pi} \bar{u}_l i\gamma_\lambda \gamma_5 (1 + \gamma_5)\, u_\nu + G_P \bar{u}_l \gamma_5 (1 + \gamma_5)\, u_\nu$$
$$= \left( -G_A \frac{m_l}{m_\pi} + G_P \right) \bar{u}_l (1 + \gamma_5)\, u_\nu . \tag{6.2.14}$$

We must then evaluate $\sum_{i,f} |T_{fi}|^2$ and insert normalisation and phase space terms. However, since we are interested in ratios of decay rates then it is apparent that $\sum_{i,f} \bar{u}_l (1 + \gamma_5) u_\nu^2$ will approximately cancel and also the phase space factors‡, hence rough limits on the branching ratio are

$$\frac{\Gamma(\pi \to e)}{\Gamma(\pi \to \mu)} \sim \left| \frac{G_A m_e - G_P m_\pi}{G_A m_\mu - G_P m_\pi} \right|^2 \sim \begin{matrix} 10^{-4} & G_P = 0 \\ 1 & G_A = 0. \end{matrix}$$

Exact calculation gives excellent agreement with experiment for $G_P = 0$.

This situation repeats itself throughout the whole range of leptonic interactions—all of them appear to be V or A or a mixture of the two.

### 6.2(d). Selection rules for leptonic decays

With the exception of muon decay all the known leptonic decays involve hadrons in the initial state. The observed features of all these processes

‡ An exact evaluation of the traces of the $\gamma$-matrices and the branching ratio is given in Appendix A6.

can be successfully represented by the matrix element

$$T_{fi} = \langle H'|\, V_\lambda - A_\lambda \,|H\rangle\, \bar{u}_e \gamma_\lambda (1 + \gamma_5)\, u_\nu \qquad (6.2.15)$$

where we have restricted outselves to $e^-\bar{\nu}$ final states. In this expression $H$ and $H'$ refer to initial and final hadronic states respectively, whilst $V_\lambda$ and $A_\lambda$ represent operators with the transformation properties of vectors and axial vectors [the implications of the sign will be discussed in § 6.3(b)]. The state $\langle H'|$ can, of course, be $\langle 0|$ as in the example of $\pi$ decay discussed in the previous section, also $V_\lambda$ and $A_\lambda$ are not necessarily both present.

In Chapter 1 we indicated that isospin conservation was violated in weak interactions and also that strangeness was broken for the strange particles. The point we wish to illustrate in the present section is that the violations occur in a regular way.

Consider firstly examples of leptonic decay which do not involve a change in strangeness

| | $n \to p\, e^-\bar{\nu}$ | | $\Sigma^- \to \Lambda\, e^-\bar{\nu}$ | | $\pi^+ \to e^+\nu$ |
|---|---|---|---|---|---|
| $S$ | 0 | 0 | $-1$ | $-1$ | 0 |
| $I_3$ | $-\frac{1}{2}$ | $\frac{1}{2}$ | $-1$ | 0 | 1 |
| $I$ | $\frac{1}{2}$ | $\frac{1}{2}$ | 1 | 0 | 1 |

An inspection of these reactions suggests the law $\Delta S = 0$, $|\Delta I| = 1$ for the hadronic part of the matrix element; all the other known interactions involving $\Delta S = 0$ appear to be consistent with this law.

Next consider the decays involving a change in strangeness. They appear to be limited to $|\Delta S| = 1$. The absence of $|\Delta S| = 2$ transitions follows from the failure to observe $\Xi$ decays of the type

| | $\Xi^0 \to p\, e^-\bar{\nu}$ | | $\Xi^0 \to p\, \pi^-$ | | |
|---|---|---|---|---|---|
| $S$ | $-2$ | 0 | $-2$ | 0 0. | (6.2.16) |

The upper limit on the branching ratios for these processes is $\sim 10^{-3}$. Since the branching ratio for the $|\Delta S| = 1$ leptonic decay of the $\Xi$

$$\Xi^- \to \Lambda\, e^-\bar{\nu}$$

is $\sim 10^{-3}$, the limit on the left-hand reaction in (6.2.16) is obviously a weak one. Indirect evidence for the absence of $|\Delta S| = 2$ transitions follows from the mass difference of the $K_1^0$ and $K_2^0$ mesons (see *P.E.P.*, p. 576).

Now consider the implications of the $|\Delta S| = 1$ rule for the Gell-Mann, Nishijima relation (1.4.3),

$$\frac{Q}{e} = I_3 + \frac{1}{2}(B + S).$$

If we define the change in charge as $\Delta Q = (Q/e)_{\text{final}} - (Q/e)_{\text{initial}}$ then the above relation yields for the hadronic part of the transition

$$\Delta Q = \Delta I_3 + \frac{1}{2}\Delta S$$

since $B$ is conserved in weak interactions. We then find

$$\begin{aligned} \Delta Q = \Delta S &\rightarrow |\Delta I_3| = \frac{1}{2} \rightarrow |\Delta I| \geqq \frac{1}{2} \\ \Delta Q = -\Delta S &\rightarrow |\Delta I_3| = \frac{3}{2} \rightarrow |\Delta I| \geqq \frac{3}{2}. \end{aligned} \tag{6.2.17}$$

Thus $|\Delta I| = \frac{1}{2}$ implies $\Delta Q = \Delta S$ but not vice versa.

All the available experimental data (Filthuth, 1969) is consistent with dominant $\Delta Q = \Delta S$ and $\Delta I = \frac{1}{2}$ transitions, for example

$$\frac{\Gamma(\Sigma^+ \rightarrow ne^+\nu)}{\Gamma(\Sigma^- \rightarrow ne^-\bar{\nu})} \leqq 0{\cdot}4 \times 10^{-2}.$$

Nevertheless it is always difficult to detect a weak amplitude in the presence of a stronger one and so $\Delta Q = -\Delta S$, $|\Delta I_3| \leqq \frac{3}{2}$ interactions cannot be dogmatically ruled out.

The rule $\Delta Q = \Delta S$ has no meaning for the decay of strange particles to purely hadronic final states, as an inspection of typical decay processes soon reveals. A dominating $\Delta I = \frac{1}{2}$ transition occurs (Filthuth, 1969), but the well-established reaction $K^+ \rightarrow \pi^+\pi^0$ involves an $I = 2$ final state‡ and so must lead to $\Delta I \geqq \frac{3}{2}$.

One interesting facet of the $\beta$-decay reactions is the smallness of the branching ratios for $|\Delta S| = 1$ processes, for example

$$\frac{\Gamma(\Lambda \rightarrow pe^-\bar{\nu})}{\Gamma_\Lambda} \begin{array}{ll} \sim 10^{-4} & \text{experiment} \\ \sim 10^{-3} & \text{calculation} \end{array}$$

where $\Gamma_\Lambda$ is the total decay width (measured) and the calculations for the numerator use the same matrix element and coupling strengths as in nuclear $\beta$-decay [§ 6.2(b)]. An explanation for this discrepancy has been given by Cabbibo (1963) within the framework of $SU_3$. Cabbibo suggested that the internal axes for a pure weak interaction space did not coincide with those for hadronic space and the difference could be given a geometrical parameterisation. A universal weak interaction for leptonic decay could then the represented by a Hamiltonian

$$\mathscr{H}_I = \frac{G_\mu}{\sqrt{2}}(j_\lambda + \mathrm{H}_\lambda^C \cos\vartheta + \mathrm{H}^{NC} \sin\vartheta)\, j_\lambda^\dagger + \text{Hermitian conjugate} \tag{6.2.18}$$

and the individual matrix elements then arise through

$$S_{fi} = -i\langle f| \int d^4x\, \mathscr{H}_I\, |i\rangle$$

[compare (4.1.9)]. In equation (6.2.18) $\mathrm{H}^C$ and $\mathrm{H}^{NC}$ are the $\Delta S = 0$ and 1 hadronic parts of the interaction respectively in field operator notation [§ 3.6 and 4.1(b)], $G_\mu$ the muon decay strength and

$$j_\lambda^\dagger = \bar{\psi}_e\gamma_\lambda(1 + \gamma_5)\,\psi_{\nu_e} + \bar{\psi}_\mu\gamma_\lambda(1 + \gamma_5)\,\psi_{\nu_\mu}.$$

A remarkably good fit to all the leptonic decay rates can be obtained with $\vartheta = 0{\cdot}235 \pm 0{\cdot}006$ (Filthuth, 1969). Consistency between the decay rates for nuclear $\beta$-decay and muon decay is found, and typical branching ratios for hyperon decay are given in Table 6.1.

‡ The $\pi^+\pi^0$ combination has $j^P = 0^+$, $I_3 = +1$; an $I = 1$ state is forbidden by the requirements of Bose–Einstein symmetry, hence $I = 2$ and $\Delta I \geqq \frac{3}{2}$. It is worth noting that $\Gamma(K^+ \rightarrow \pi^+\pi^0) \sim 10^{-2}\,\Gamma(K^0 \rightarrow \pi^+\pi^-)$ thus indicating that the $\Delta I = \frac{3}{2}$ transition is much weaker than the $\Delta I = \frac{1}{2}$ transition.

TABLE 6.1

| Decay | Measured rate | Calculated rate |
|---|---|---|
| $\Sigma^- \rightarrow \Lambda e^- \bar{\nu}$ | $(0.59 \pm 0.06) \times 10^{-4}$ | $0.62 \times 10^{-4}$ |
| $\Lambda \rightarrow p e^- \bar{\nu}$ | $(0.83 \pm 0.08) \times 10^{-4}$ | $0.86 \times 10^{-3}$ |
| $\Sigma^- \rightarrow n e^- \nu$ | $(1.10 \pm 0.05) \times 10^{-4}$ | $1.01 \times 10^{-4}$ |

## 6.3. Problems in Weak Interactions

### 6.3(a). Intermediate vector bosons

The V, A theory of weak interactions can be expressed in its simplest form in $\mu$-decay (6.2.13)

$$T_{fi} = \frac{G_\mu}{\sqrt{2}} \bar{u}_\nu \gamma_\lambda (1 + \gamma_5) u_\mu \bar{u}_e \gamma_\lambda (1 + \gamma_5) u_\nu .$$

Since this interaction involves no hadrons it can be regarded as a point interaction (Fig. 6.5) and $G_\mu$ as a constant. The expression appears to

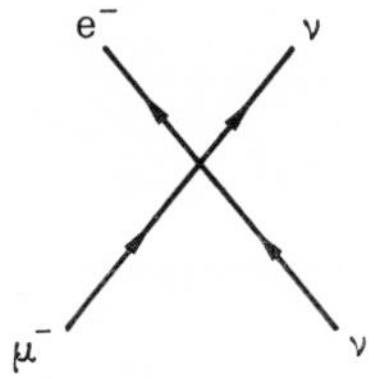


FIG. 6.5.

describe all $\mu$-decay data exactly. However, it should also work for

$$\mu^- \bar{\nu} \rightarrow e^- \bar{\nu}$$

and, if we neglect lepton masses, it is not difficult to show that at high energies the total cross-section becomes‡

$$\sigma_T = \frac{4}{\pi} G_\mu^2 p_c^2 \tag{6.3.1}$$

‡ The gross features of this behaviour are apparent from equation (2.5.10) if we set $E_\nu \sim E_n \sim p_c$.

where $p_c$ is the momentum in the $c$-system. Now since this is a point interaction it should also be an $s$-wave interaction and this requirement gives [§ 7.1 (b)]

$$\sigma_T < \frac{\pi}{2p_c^2}.$$

Thus at very high energies the theory must break down.

One way out of this dilemma is to postulate the existence of a massive vector boson (Fig. 6.6)

$$T_{fi} = G_W^2 \bar{u}_\nu \gamma_\lambda (1 + \gamma_5) u_\mu \frac{1}{q^2 + M_W^2} \left( \delta_{\lambda\sigma} + \frac{q_\lambda q_\sigma}{M_W^2} \right) \bar{u}_e \gamma_\sigma (1 + \gamma_5) u_\nu . \tag{6.3.2}$$

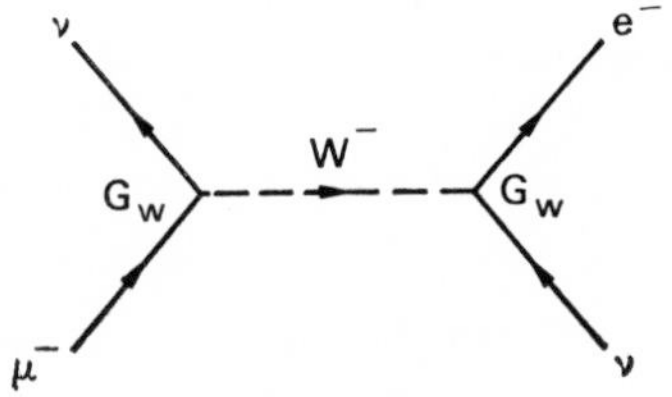


FIG. 6.6.

Thus for $q^2 \ll M_W^2$

$$\frac{G_W^2}{M_W^2} = \frac{G_\mu}{\sqrt{2}}$$

and the effects are indistinguishable. For $q^2 \gg M_W^2$ the cross-section falls as $\sim 1/p_c^2$.

Considerable effort has been made to establish the existence of the boson. The first attempt was a search for the process $\mu \to e\gamma$. As a first estimate one might expect a branching ratio (compare Fig. 6.7)

$$\frac{\Gamma(\mu \to e\gamma)}{\Gamma(\mu \to e\nu\bar{\nu})} \sim e^2 \sim \alpha = \frac{1}{137}.$$

More detailed estimates give $\sim 10^{-4}$. Experimentally the ratio is $< 10^{-8}$. This result does not necessarily mean that the vector bosons do not exist, as it turns out that the neutrinos associated with a muon vertex are not the same as those associated with an electron vertex (see below) and so the diagram is not admissible.

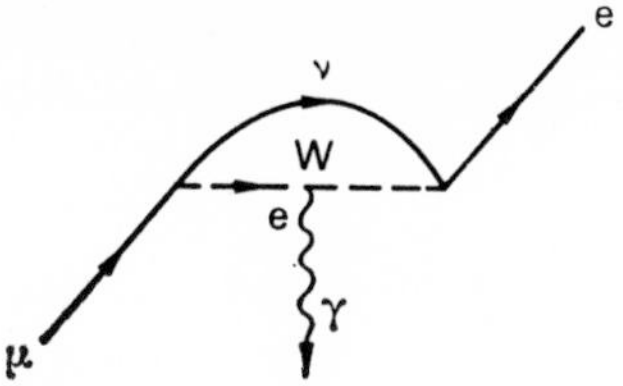


FIG. 6.7.

The search for different types of neutrinos and massive vector bosons has been made in high energy neutrino experiments. The experiments have yielded the following results:

1. $\nu_e \neq \nu_\mu$

The neutrinos are generated in $\pi$-decay

$$\pi \to \mu\nu_\mu$$

and subsequently undergo quasi-elastic interactions in suitable targets; there muons are found but not electrons

$$\nu_\mu N \to N'\mu$$
$$\not\to N'e.$$

Thus the absence of electrons implies the existence of a "muonic conservation law". These experiments have also shown that the nucleon has a weak form factor similar to that found in electromagnetic interactions [compare § 5.3(a)]

$$G(q^2) \sim \frac{1}{(1 + q^2/0{\cdot}71)^2}.$$

2. $\nu_\mu \neq \bar{\nu}_\mu$

The experiments show that neutrinos from $\pi^+$-decay cause $\mu^-$ production and vice versa. This result is easily explained in terms of the concept of the conservation of a leptonic charge $l_\mu$—"conservation of leptons"

$$\begin{array}{lcc} & \pi^+ \to \mu^+\nu_\mu & \nu_\mu N \to N'\mu^- \\ l_\mu & +1-1 & -1 \qquad\quad -1 \\ & \pi^- \to \mu^-\bar{\nu}_\mu & \bar{\nu}_\mu N \to N'\mu^+ \\ l_\mu & -1+1 & +1 \qquad\quad +1. \end{array}$$

A similar conclusion regarding lepton conservation has been reached for the neutrinos associated with $\beta$-decay.

3. $M_W > 1{\cdot}5$ GeV

The boson could manifest itself through the reaction displayed in Fig. 6.8. A pair of muons of opposite sign would appear. The incident neutrinos have a continuous energy spectrum which falls rapidly with increasing energy. The absence of muon pairs allows the lower limit to be placed on $M_W$. Thus no positive evidence for the existence of the heavy vector boson has yet been obtained.

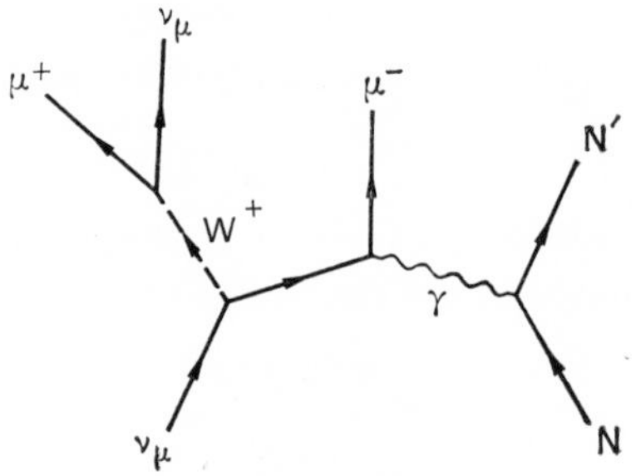


FIG. 6.8.

### 6.3(b). *CP* and the $K^0$, $\bar{K}^0$ system

Ample evidence exists that *P* and *C* violation occurs in weak interactions. So far we have not mentioned *T* violation, nor have we considered possible combinations of *C*, *P* and *T*.

An important theorem ("the *CPT* theorem") states that if a system is invariant under the combination *CPT* taken in any order then the masses and lifetimes of particles are the same (*P.E.P.*, p. 369). All the experimental evidence supports this requirement. Overall invariance under *CPT* does not preclude, however, violations of two of the three operations. Thus we can have violations of *C* and *P* separately but not necessarily of *CP*. With one exception all the available experimental evidence points to the conclusion that the simultaneous transformation of particles to antiparticles and the reversal of helicities leaves all the amplitudes of weakly interacting processes unchanged. One can also separately examine their behaviour under time reversal and no apparent $T$ violation is observed.

We shall now examine the latter point a little more closely. In § 6.2(d) it was stated that the observed properties of leptonic decays could be successfully represented by the matrix element (6.2.15)

$$T_{fi} = \langle H'| V_\lambda - A_\lambda |H\rangle \bar{u}_e \gamma_\lambda (1 + \gamma_5) u_l .$$

A more general operator in this equation could be obtained by including a phase factor $\phi$

$$V_\lambda + e^{i\Phi} A_\lambda .$$

A phase factor different from $\phi = 0, \pi$ would then lead to a non-zero expectation of correlations between the initial hadron polarisation and the decay plane of $e^-\bar{\nu}$, i.e. a term $\langle \mathbf{S}_N \rangle \cdot \mathbf{p}_e \times \mathbf{p}_\nu$. But an expression of this type violates time reversal [compare § 6.1(c) and (6.2.5)]

$$\langle \mathbf{S}_N \rangle \cdot \mathbf{p}_e \times \mathbf{p}_\nu \xrightarrow{T} -\langle \mathbf{S}_N \rangle \cdot \mathbf{p}_e \times \mathbf{p}_\nu .$$

At present the best measurements of $\phi$ yield

$$n \to p\, e^- \bar{\nu} \qquad \phi = 178{\cdot}7 \pm 1{\cdot}3^0$$

$$Ne^{19} \to F^{19}\, e^+ \nu \qquad \phi = 179{\cdot}6 \pm 2{\cdot}0^0$$

(report by Steinberger, 1969). In non-leptonic decays the correlation $\langle \mathbf{S}_\Lambda \rangle \cdot \langle \mathbf{S}_p \rangle \times \mathbf{p}_p$ has been examined in the process $\Lambda \to p\pi^-$ with the result $\phi = 2{\cdot}8 \pm 4^0$ (Steinberger, 1969).

Thus to within the present limits of experimental accuracy no evidence exists for $CP$ or $T$ violations in weak interactions—with one important exception. This exception occurs in the $K^0$, $\bar{K}^0$ system. The $K^0$ and $\bar{K}^0$ mesons are not eigenstates of $CP$

$$CP\,|K^0\rangle = \zeta\,|\bar{K}^0\rangle .$$

However, the pion states $\pi^+ \pi^-$ into which they decay are eigenstates of $CP$; the equations (6.1.5) and (6.1.8) lead to the result

$$CP\,|\pi^+\pi^-\rangle = |\pi^+\pi^-\rangle . \tag{6.3.3}$$

This observation prompted Gell-Mann and Pais (1955) to suggest that the systems which were decaying represented linear combinations of $K^0$ and $\bar{K}^0$

$$|K_1^0\rangle = \frac{1}{\sqrt{2}}(|K^0\rangle + |\bar{K}^0\rangle) \qquad |K_2^0\rangle = \frac{1}{\sqrt{2}}(|K^0\rangle - |\bar{K}^0\rangle)$$

then

$$CP\,|K_1^0\rangle = |K_1^0\rangle \qquad CP\,|K_2^0\rangle = -|K_2^0\rangle.$$

Now systems of three pions $|\pi^+\pi^-\pi^0\rangle$ can be easily shown to form negative eigenstates of $CP$ [compare § 7.1 (e)] but because of phase space considerations their decay time would be expected to be much longer than for $2\pi$ decay. At the time of the suggestion of Gell-Mann and Pais only the $2\pi$ decay mode was known. A search for the long lived $K_2^0$ system was made and it was duly found. However, in 1964 the work of Christenson *et al.* revealed a weak decay mode of $K_2^0$ to $2\pi$

$$\frac{K_2^0 \to \pi^+\pi^-}{K_2^0 \to \text{all modes}} \sim 2 \times 10^{-3}.$$

Thus $CP$ violation occurs in the $K^0, \bar{K}^0$ system, but in contrast to maximal $C$ and $P$ violation in weak interactions $CP$ violation is minimal. No evidence for its occurrence has been found outside the $K^0, \bar{K}^0$ system and no satisfactory explanation exists for its occurrence. The elucidation of the details of the $K^0, \bar{K}^0$ decay amplitude is complicated to explain and will not be attempted here. Because of the smallness of the $CP$ violating amplitude the associated experiments which attempt to establish the form of the amplitude are difficult and often appear to lead to contradictory results.

CHAPTER 7

# STRONG INTERACTIONS

QUANTUM electrodynamics provides exact agreement between theory and experiment because it is based upon a parameter $\alpha \sim 1/137$, hence the calculation of higher-order terms in an electromagnetic process can be made to rapidly converge. No similar situation exists in strong interactions, since the natural development parameter is of order 10. One therefore is forced to proceed in an empirical fashion and exploit to the maximum general invariance principles.

## 7.1. Quantum Numbers of the Resonances

### 7.1(a). Introduction

One of the distinguishing features of the strong interactions is the large number of particles or resonances, and these must all be classified and if possible coordinated. We therefore start by outlining methods to determine their spins and parities.

Resonances are normally formed in one of two types of experiment—formation and production (Fig. 7.1). In the former a cross-section is measured as a function of energy and peaks are sought; in the latter mass distributions in multiparticle final states are examined.

The formation experiments show that at low energies the resonances dominate the cross-section for $\pi^{\pm}N$ and $K^{-}N$ interactions, but no resonances are found in $K^{+}N$ or $NN$ reactions. In general the properties of resonances found in formation experiments are determined by phase shift analyses, and so we first examine this technique.

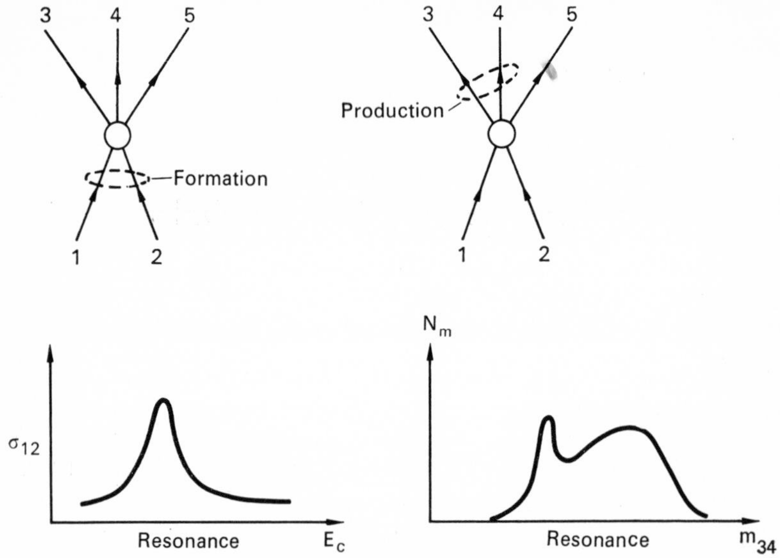


FIG. 7.1. Principles of formation and production experiments to detect resonances.

### 7.1(b). Elastic scattering and phase shifts

Consider the elastic scattering of spinless particles; we may define a scattering amplitude $f(\theta)$ by the following relation [compare (2.4.3)]

$$\frac{d\sigma}{d\Omega} = |f(\theta)|^2 = \frac{1}{(2\pi n E_c)^2}|T_{fi}|^2$$

and so to within a phase factor

$$f(\theta) = \frac{1}{2\pi n E_c} T_{fi}. \tag{7.1.1}$$

Let the particles have momentum $k = |\mathbf{k}|$ in the $c$-system and the scattered particles move in the direction $\theta$, $\phi$, then

$$T_{fi} = \langle f|\,T\,|i\rangle \qquad |f\rangle = |k\theta\phi\rangle \equiv |\theta\phi\rangle$$

$$|i\rangle = |k00\rangle \equiv |00\rangle.$$

We wish to expand this amplitude into states of angular momentum ("partial waves"). This result can be achieved with the aid of the completeness relation (Appendix A3)

$$T_{fl} = \sum_{l'm'} \sum_{lm} \langle \theta\phi | \, l'm' \rangle \, \langle l'm' | \, T \, | lm \rangle \, \langle lm \, | \, 00 \rangle . \qquad (7.1.2)$$

If we quantise along the $z$ (incident particle) axis

$$m = 0 \quad \text{because} \quad m \equiv m_z = yk_x - xk_y = 0$$

since $k_x = k_y = 0$; also conservation of angular momentum gives

$$m = m' = 0 \qquad l = l'$$

thus (Appendix A3)

$$\langle \theta\phi \, | \, l0 \rangle = Y_l^{0*}(\theta, \phi) = \sqrt{\frac{2l + 1}{4\pi}} \, P_l^0(\theta) \qquad (7.1.3)$$

$$\langle l0 \, | \, 00 \rangle = Y_l^0(0, 0) = \sqrt{\frac{2l + 1}{4\pi}}$$

$$T_{fl} = \sum_l \frac{(2l + 1)}{4\pi} T_l P_l^0(\theta) \qquad (7.1.4)$$

Thus $T_l$ gives us the scattering amplitude for the $l$th partial wave; we may parameterise it as

$$T_l = 8\pi^2 E_c n \frac{\eta_l - 1}{2ik} \qquad (7.1.5)$$

and hence obtain the Rayleigh scattering formula

$$f(\theta) = \sum_l (2l + 1) \frac{\eta_l - 1}{2ik} P_l^0(\theta) \qquad (7.1.6)$$

then

$$\frac{d\sigma}{d\Omega} = |f(\theta)|^2 = \frac{1}{4k^2} \sum_l |(2l + 1)(\eta_l - 1) \, P_l^0(\theta)|^2 \qquad (7.1.7)$$

Since

$$\int d\Omega \, P_{l'} P_l = \frac{4\pi}{2l + 1} \delta_{l'l}$$

the total scattering cross-section $\sigma_{sc}$ is

$$\sigma_{sc} = \int d\Omega\, |f(\theta)|^2 = \frac{\pi}{k^2} \sum_l (2l+1)\, |\eta_l - 1|^2. \tag{7.1.8}$$

This is not necessarily the total cross-section as inelastic processes can also occur. The total cross-section can be obtained by the optical theorem. We previously showed that the unitarity requirement for the $S$ operator (4.3.5) implied that

$$2 \operatorname{Im} T_{fi} = \sum_n \frac{(2\pi)^4}{N_n^2} \delta(p_n - p_i) \langle f|\, T\, |n\rangle \langle n|\, T^\dagger\, |i\rangle$$

where $n$ represented possible intermediate states. Now if we consider forward scattering we have $i = f$ and so

$$2 \operatorname{Im} T_{ii} = \sum_n \frac{(2\pi)^4}{N_n^2} \delta(p_n - p_i)\, |T_{ni}|^2.$$

But $\sum_n$ represents a sum of all systems compatible with $i$, and so we can replace $n$ with $f$—the symbol for final states

$$2 \operatorname{Im} T_{ii} = \sum_f \frac{(2\pi)^4}{N_f^2} |T_{fi}|^2\, \delta(p_f - p_i)$$

and if we use equations (2.3.4), (2.4.4) and (2.3.11) and multiply the right side by

$$1 = \frac{kE_c n_i^2}{kE_c n_i^2} = \frac{kE_c n_i^2}{kE_c} \frac{E_a E_b}{N_i^2} = kE_c n_i^2 \frac{1}{v_{ab} N_i^2}$$

where $a$ and $b$ are labels for the initial particles and $v_{ab}$ is their relative velocity, we obtain

$$2 \operatorname{Im} T_{ii} = kE_c n_i^2 \frac{1}{v_{ab} N_i^2} \sum_f \frac{(2\pi)^4}{N_f^2} |T_{fi}|^2\, \delta(p_f - p_i)$$

$$= kE_c n_i^2 \sigma_T$$

where $\sigma_T$ is the total cross-section [compare (2.3.5)]. The insertion of equation (7.1.1) then gives us the optical theorem

$$\begin{aligned}\sigma_T &= \frac{2}{k}\frac{1}{E_c n_i^2}\,\mathrm{Im}\,T_{ii}\\ &= \frac{2}{k}\frac{1}{E_c n_i^2}\,2\pi E_c n\,\mathrm{Im}\,f(0)\\ &= \frac{4\pi}{k}\,\mathrm{Im}\,f(0)\end{aligned} \tag{7.1.9}$$

since $n_i^2 = n$ for identical particles in the initial and final states.

Since (7.1.6)

$$f(\theta) = \sum_l (2l+1)\frac{(\eta_l - 1)}{2ik}P_l^0(\theta)$$

and $P_l^0(0) = 1$, the above relation gives us

$$\sigma_T = \frac{2\pi}{k^2}\sum_l (2l+1)(1 - \mathrm{Re}\,\eta_l).$$

The difference between $\sigma_T$ and $\sigma_{sc}$ gives the reaction cross-section, i.e. the non-elastic part

$$\begin{aligned}\sigma_T - \sigma_{sc} = \sigma_r &= \frac{\pi}{k^2}\sum_l (2l+1)[2 - 2\,\mathrm{Re}\,\eta_l - |1 - \eta_l|^2]\\ &= \frac{\pi}{k^2}\sum_l (2l+1)(1 - |\eta_l|^2).\end{aligned} \tag{7.1.10}$$

Since we cannot have negative cross-sections the above equations place certein restrictions on $\eta_1$ for each partial wave

$$\begin{aligned}\sigma_r &\geqq 0 \qquad |\eta_l|^2 \leqq 1\\ \sigma_T &\geqq 0 \qquad \mathrm{Re}\,\eta_l \leqq 1\\ \sigma_{sc} &\geqq 0 \qquad 1 + |\eta_l|^2 \geqq 2\,\mathrm{Re}\,\eta_l.\end{aligned}$$

Thus it is customary to parameterise $\eta_l$ as

$$\eta_l = \varrho_l\, e^{2i\delta_l} \qquad \varrho_l \leqq 1 \quad \delta_l \text{ real} \tag{7.1.11}$$

and an inspection of (7.1.10) shows that $\varrho_l$ indicates the degree of elasticity; $\delta_l$ is called the phase shift. The scattering amplitude therefore becomes

$$f(\theta) = \sum_l (2l+1)\frac{\varrho_l e^{2i\delta_l} - 1}{2ik} P_l^0(\theta). \tag{7.1.12}$$

If we represent the centre part of the above equation as $T_e$, then in the elastic limit $\varrho_l \to 1$

$$T_e = \frac{\varrho_l e^{2i\delta_l} - 1}{2i} \underset{\varrho_l \to 1}{=} \frac{e^{2i\delta_l} - 1}{2i} = \frac{i}{2} + \frac{1}{2}(\sin 2\delta_l - i \cos 2\delta_l) \tag{7.1.13}$$

This is the equation of a circle of radius $\frac{1}{2}$ and centre $i/2$ and is represented by Fig. 7.2. The circle is called the unitarity circle. If $\varrho_l < 1$ we can write

$$T_e = \frac{\varrho e^{2i\delta} - 1}{2i} = \frac{\varrho(e^{2i\delta} - 1)}{2i} + \frac{\varrho - 1}{2i},$$

i.e. a circle of reduced radius inside the unitarity circle and with displaced centre.

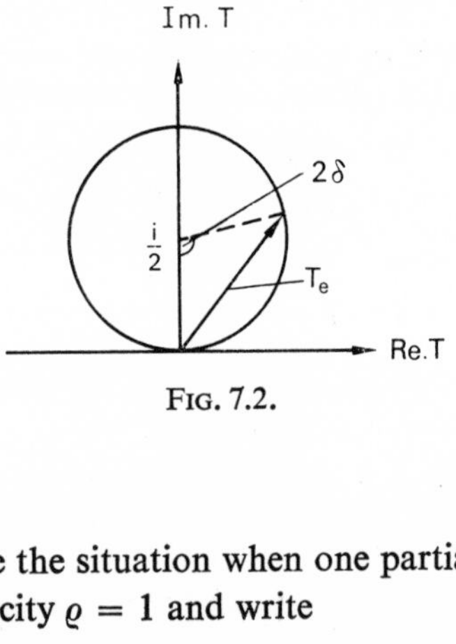


FIG. 7.2.

### 7.1(c). Resonances

Now let us examine the situation when one partial wave resonates. We assume for simplicity $\varrho = 1$ and write

$$T_e = \frac{e^{2i\delta} - 1}{2i} = e^{i\delta}\frac{(e^{i\delta} - e^{-i\delta})}{2i}$$

$$= \frac{\sin\delta}{e^{-i\delta}} = \frac{1}{\cot\delta - i}. \tag{7.1.14}$$

It is evident that

1. a peak (i.e. a resonance) appears in the cross-section as $T_e$ passes through $\cot\delta = 0$, i.e. $\delta = (n + \frac{1}{2})\pi$; this point corresponds to the top of the unitarity circle (Fig. 7.2).
2. $T_e$ is purely imaginary at this point.

In practice life is more complicated than this. One often starts with a purely elastic partial wave at low energies which becomes progressively more inelastic as the energy increases. This situation is illustrated in Fig. 7.3.

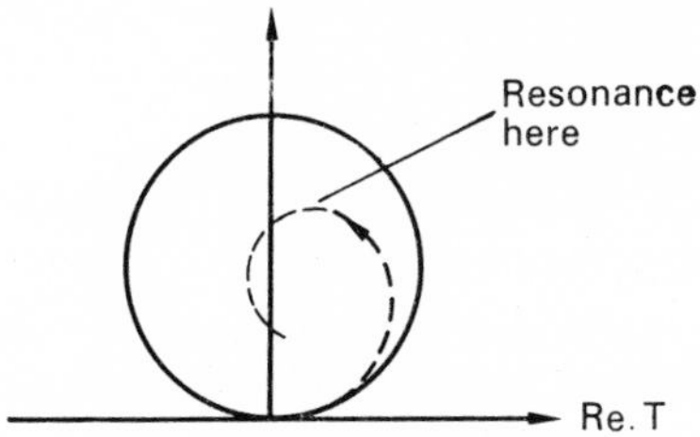


FIG. 7.3.

The term $\cot\delta$ can be related to the parameters of a particle resonance; considering Fig. 7.4, the Feynman rules (§ 4.4) give the following amplitude for an intermediate state in the $s$ channel ($m_R$ is the mass of the resonance and $\Gamma$ its width)

$$T_{fi} \equiv \frac{1}{k^2 + \left(m_R - i\dfrac{\Gamma}{2}\right)^2}$$

and since $k^2 = -E_c^2 = -s$ (4.3.1; 4.3.2)

$$T_{fi} \equiv \frac{1}{\dfrac{m_R^2 - s}{m_R\Gamma} - i}$$

where we assume $\Gamma \ll m_R$. A comparison with equation (7.1.14) then suggests

$$\cot\delta = \frac{m_R^2 - s}{m_R\Gamma}. \tag{7.1.15}$$

Thus if one examines the amplitude $T_e$ as a function of energy it reaches a maximum and then decreases (Fig. 7.1), and the sharpness of the peak in $|T_e|^2$ depends on the width $\Gamma$.

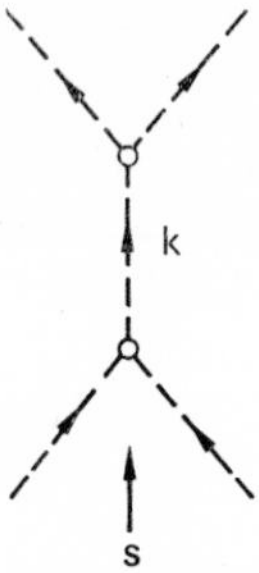


FIG. 7.4.

It is worth noting the relation between a given partial wave and a Feynman diagram in detail as it gives us a clue on how to write Feynman amplitudes for systems of high spin. Consider the situation $\pi\pi \to \varrho \to \pi\pi$; equations (7.1.12) and (7.1.14) give us

$$f(\theta) = \sum_l (2l+1) \frac{1}{k(\cot \delta_l - i)} P_l^0(\theta)$$

Since the $\varrho$ has spin $j \equiv l = 1$, we then have from (7.1.1) (with $n = 4$)

$$T_{fi} = 2\pi E_c n f(\theta) = 8\pi E_c \frac{3 \cos \vartheta}{k(\cot \delta - i)}\cdot$$

On the other hand, the Feynman rules (4.3.11) give a matrix element

$$T_{fi} = \frac{\gamma^2(\mathbf{k}_3 - \mathbf{k}_4)\cdot(\mathbf{k}_1 - \mathbf{k}_2)}{m_\varrho \Gamma\left(\dfrac{m_\varrho^2 - s}{m_\varrho \Gamma} - i\right)}$$

where the evaluation has been made in the $c$-system, and since [(5.3.6) and (5.3.7)]

$$\mathbf{k}_1 = -\mathbf{k}_2 \qquad \mathbf{k}_3 = -\mathbf{k}_4$$

$$|\mathbf{k}_1| = |\mathbf{k}_2| = |\mathbf{k}_3| = |\mathbf{k}_4| = k$$

$$\Gamma = \frac{\gamma^2}{4\pi}\frac{2k_\varrho^3}{3m_\varrho^2} \quad \text{where} \quad m_\varrho^2 = 4(m_\pi^2 + k_\varrho^2)$$

then

$$T_{fi} = 8\pi m_\varrho \frac{3k^2 \cos \vartheta}{k_\varrho^3 \left( \dfrac{m_\varrho^2 - s}{m_\varrho \Gamma} - i \right)} \tag{7.1.16}$$

Thus the two amplitudes are indistinguishable in the neighbourhood of $s = m_\varrho^2$. It is evident from the above comparison that a good guess for the propagator in the $s$-channel for a system of spin $j$, in the rest system of $j$, is (neglecting $\Gamma^2/4$)

$$\text{propagator} = \frac{g^2 P_j(\theta)}{k^2 + \left( m_j - i \dfrac{\Gamma}{2} \right)^2} = \frac{g^2 P_j(\theta)}{(m_j^2 - s) - i m_j \Gamma} \tag{7.1.17}$$

where $g$ is a coupling strength and $m_j$ is the mass of the particle $j$.

### 7.1(d). Complications

In practice not many systems have spinless particles in the initial and final states. However, partial wave analysis has been used to establish the spin and parity of the $\varrho$ and $f$ mesons by exploiting situations similar to that illustrated in Fig. 7.5. In this situation the meson propagator must have low four momentum ($q$) so that the exchanged meson is nearly on the mass shell, i.e. $q^2 \to -m_\pi^2$.

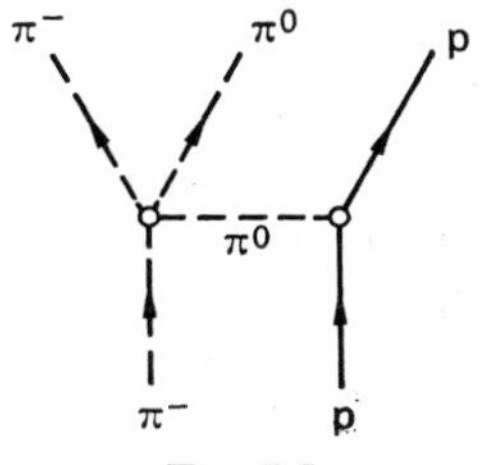


FIG. 7.5.

Phase shift analysis has been used extensively in determining the quantum numbers of the baryon resonances. These are formed from $\pi N$ and $K^- N$ initial states and so allowance must be made for the complications of nucleon spin. This can be done quite easily with the aid of

Clebsch–Gordan coefficients. Suppose we have a pure angular momentum state $|j_1m_1; j_2m_2\rangle$; this can be expanded in terms of a complete set of orthogonal bases of states $|JM\rangle$ (Appendix A3)

$$
\begin{aligned}
|j_1m_1; j_2m_2\rangle &= \sum_{J,M} |JM\rangle \langle JM | j_1m_1; j_2m_2\rangle \\
&= \sum_{J,M} C_{J\ j_1\ j_2}^{M\ m_1\ m_2} |JM\rangle \\
&= \sum_{J} C_{J\ j_1\ j_2}^{M\ m_1\ m_2} |JM\rangle
\end{aligned}
\tag{7.1.18}
$$

since conservation of angular momentum limits us to $m_1 + m_2 = M$. The terms $C$ are called Clebsch–Gordan coefficients; their evaluation is straightforward (see *P.E.P.*, p. 197) and tables are readily available (Appendix A3).

In handling spin terms in $T_{fi}$ we simply extend our previous completeness relation (7.1.2):

$$
\begin{aligned}
\langle f | T | i\rangle &= \langle \theta, \phi, s' | T | 0, 0, s\rangle \qquad s = \text{initial spin orientations} \\
&\qquad\qquad\qquad\qquad\qquad\quad\; s' = \text{final spin orientations} \\
&= \sum_{\substack{lm\\ l'm'\\ JM}} \langle \theta, \phi, s' | l'm's'\rangle \langle l'm's' | JMl'\rangle \langle JMl' | T | JMl\rangle \\
&\qquad\qquad \times \langle JMl | lms\rangle \langle lms | 0, 0, s\rangle \\
&= \sum_{J,l',l} \chi_f^{\dagger} Y_{l'}^{m'*}(\theta, \phi)\, C_{Jl's_f}^{sm's'} \langle Jsl' | T | Jsl\rangle\, C_{Jls_i}^{s0s} Y_l^0(0, 0)\, \chi_i
\end{aligned}
\tag{7.1.19}
$$

where $\chi_i$ and $\chi_f$ are spin functions for the initial and final states respectively. The reduction of equation (7.1.19) to a useable form is straightforward but lengthy. The scattering of particles with spin 0 and $\frac{1}{2}$ ($\pi N$ for example) is given on p. 244 of *P.E.P.*—an alternative method is given in Appendix A3. The result is a combination of two amplitudes—a spin flip $h(\theta)$ and non-spin flip $g(\theta)$

$$
g(\theta) = \frac{1}{k} \sum_{l=0}^{\infty} [(l + 1)\, T_{l+} + lT_{l-}]\, P_l^0(\theta)
$$

$$
h(\theta) = \frac{1}{k} \sum_{l=0}^{\infty} (T_{l+} - T_{l-})\, P_l^1(\theta)
$$

and

$$\frac{d\sigma}{d\Omega} = |g(\theta)|^2 + |h(\theta)|^2 \tag{7.1.20}$$

where $l_+ \equiv j = l + \frac{1}{2}$, $l_- \equiv j = l - \frac{1}{2}$. The $T$ functions are as in (7.1.13) with appropriate subscripts.

It is apparent that equation (7.1.20) can lead to phase ambiguities in the determination of amplitudes. Consider also the following problem. For the strong isospin 3/2 resonance with mass $m_R = 1236$ MeV (200 MeV lab kinetic energy for the incident pion) the measured angular distribution is $\sim 1 + 3\cos^2\theta$ for $\pi^{\pm}p$ scattering. Equation (7.1.20) yields

$$\frac{d\sigma}{d\Omega} = \frac{1}{k^2}(1 + 3\cos^2\theta)\,|T_{l+}|^2 \quad \text{for} \quad l = 1 \quad j = \frac{3}{2}$$

$$= \frac{1}{k^2}(1 + 3\cos^2\theta)\,|T_{l-}|^2 \qquad l = 2 \quad j = \frac{3}{2}.$$

Thus since parity $P = -(1)^l$ for $\pi N$ scattering,‡ the angular distribution gives the spin but not the parity of the resonance. Other ambiguities also exist; they can in practice all be resolved by considering scattering from polarised targets, which yield left–right assymetries proportional to Im $(gh^*)$, and from examining the interference with other partial waves of known properties.

### 7.1(e). Spin-parity analysis for production resonances

The numbers and types of particles into which a final state decays often gives us information on the quantum numbers of the state, for example a $2\pi$ decay mode limits us to $0^+$, $1^-$, $2^+$ ... [equation (6.1.6)]. Most of the information on systems which decay to two particles in production experiments has been obtained from phase shift analysis together with the assumption of particle exchange (compare Fig. 7.5).

‡ The extra minus sign arises because the pion has odd parity.

Now consider systems decaying to three particles. The decay rate is given by equations (2.3.8) and (2.3.19)

$$d\Gamma \propto \overline{\sum_i} \sum_f |T_{fi}|^2 \quad \text{phase space}$$

$$\propto \overline{\sum_i} \sum_f |T_{fi}|^2 \, dE_1 \, dE_2$$

$$\propto \overline{\sum_i} \sum_f |T_{fi}|^2 \, dT_1 \, dT_2$$

where $T_1$ and $T_2$ are kinetic energies. Now the sum of the kinetic energies is a constant, $M - m_1 - m_2 - m_3 = Q = T_1 + T_2 + T_3$, where $M$ is the mass of the decaying particle. Since the sum of the perpendiculars from any point inside an equilateral triangle is a constant, it was pointed out by Dalitz (1957) that the kinematic behaviour of a decaying particle

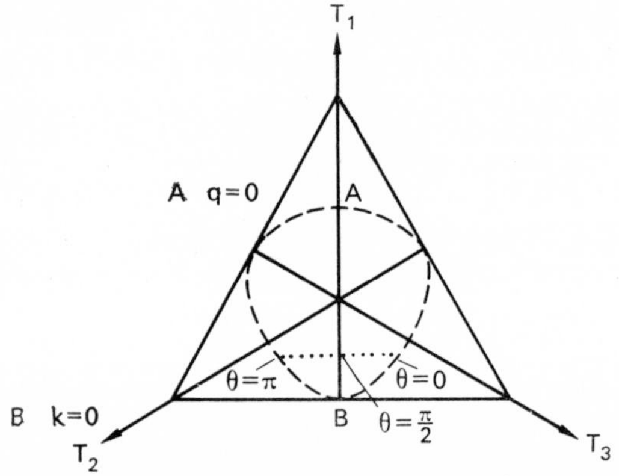


FIG. 7.6. Allowed regions for the Dalitz plot. The points $A$ and $B$ correspond to $q = 0$ and $k = 0$ respectively in Fig. 7.7. The horizontal dotted line gives the allowed values of of $T_2$ and $T_3$ for fixed $T_1$ (i.e. $k$). The dashed line traces the kinematic limits for the plot.

could be represented by a point in a triangle of height $Q$. The population of dots inside the triangle then gives $|T_{fi}|^2$; the distributions obtained in analyses of this type are known as Dalitz plots. An example for $\omega$ decay is given in Fig. 7.8.

Normally the direction of one of the particles is chosen as axis of quantisation, and the remaining two particles treated as a "diparticle" (Fig. 7.7). Then, assuming all three particles are spinless, the spin of the

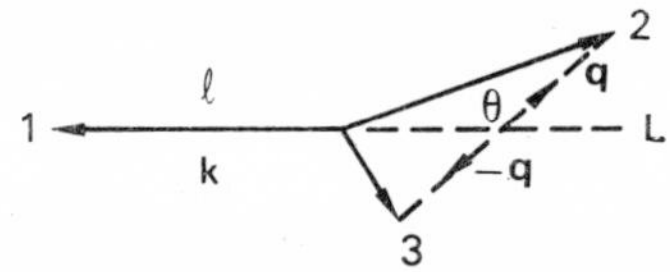


FIG. 7.7. In this figure **k** represents the momentum of particle 1 in the overall $c$-system, **q** the momentum of particle 2 in the $c$-system of the diparticle.

decaying system can be represented as $\mathbf{j} = \mathbf{L} + \mathbf{l}$. If we choose the **k** axis as direction of quantisation the final state can be represented as

$$|f\rangle \propto \sum_{l,L} C^{m\,m\,0}_{j\,L\,l} Y^0_l(0)\, Y^m_L(\theta)\, f(k, q) \qquad (7.1.21)$$

where $f$ is a scalar function of $|\mathbf{k}|$ and $|\mathbf{q}|$ and $C$ is a Clebsch–Gordan coefficient (A. 3.18). We first examine $f$ and break all our previous rules by turning $T_{fi}$ into an amplitude arising from a potential. Now a potential has spatial dependence and is an entirely nonrelativistic concept. However, we can get away with it here as one normally works in the rest system of the resonance. For simplicity we first consider a two-body final state with orbital angular momentum $l$, we then have

$$\begin{aligned} \langle f|\, T\,|i\rangle &\equiv \langle l|\, T\,|i\rangle \\ &= \sum_r \langle l|\, kr\rangle \langle kr|\, T\,|i\rangle \\ &= \sum_r j_l(kr) \langle kr|\, T\,|i\rangle \end{aligned}$$

where $j_l\,(kr)$ is the spherical Bessel function of order $l$. We insert it here by making the inspired guess that $j_l$ is the only function normally available in quantum mechanics for linking $l$, $k$ and $r$. Now a sensible range for our potential would be $R \sim 1/M$ and for

$$kR \ll l \qquad j_l(kR) \sim \frac{(kR)^l}{(2l+1)!!}$$

where $(2l + 1)!! = (2l + 1)\,(2l - 1)\,(2l - 3)\,\ldots$; thus

$$\langle f|\, T\,|i\rangle \sim \frac{(kR)^l}{(2l+1)!!} \langle kr|\, T\,|i\rangle.$$

This means that the partial waves of lowest order are normally favoured. The argument is also applicable to scattering, suppose we have a reaction

$$ab \to de$$

$$k \qquad k' \qquad c\text{-system momenta}$$

then for low momenta in the initial and final state

$$T_{fi} \propto k'^{l'} k^{l}$$

and from (2.4.3)

$$\frac{d\sigma}{d\Omega} \propto \frac{k'}{k} |T_{fi}|^2 \propto k'^{(2l'+1)}\, k^{(2l-1)}. \qquad (7.1.22)$$

Reverting to our three-particle systems we therefore represent the final state (7.1.21) as

$$|f\rangle \propto CY_l^0(0)\, Y_L^m(\theta)\, k^l q^L a(k, q) \qquad (7.1.23)$$

and (hopefully) $a$ is constant or a slowly varying scalar function of $k$ and $q$. The decay probability therefore gives

$$d\Gamma \propto |T_{fi}|^2 \propto k^{2l} q^{2L} \qquad k \to 0, q \to 0. \qquad (7.1.24)$$

Now consider possible configurations of the final systems. We assume that all three particles are pions and in general can place further restrictions on the system from symmetry considerations. Consider firstly $\tau$ decay, $\tau \equiv K^+ \to \pi^+\pi^+\pi^-$. Bose–Einstein statistics then force us to limit $L$ to 0, 2, 4 ... for $\pi^+\pi^+$, and since the intrinsic parity of three pions is $(-1)^3 = -1$, the simplest configurations are‡

| $j^P$ | $l$ | $L$ | $CY_l^0(0)\, Y_L^m(\theta)$ | $\|f\rangle$ | |
|---|---|---|---|---|---|
| $0^-$ | 0 | 0 | 1 | 1 | |
| $0^+$ | – | – | – | – | (7.1.25) |
| $1^-$ | 2 | 2 | $Y_2^1 \equiv \sin\theta\cos\theta$ | $(\mathbf{k}\times\mathbf{q})(\mathbf{k}\cdot\mathbf{q})$ | |
| $1^+$ | 1 | 0 | 1 | $\mathbf{k}$ | |

‡ Since $I_3 = +1$ the final isospin state has $I \geqq 1$ for the three pions. Isospin conservation is broken in $\tau$-decay. Note that it is impossible to construct a $0^+$ configuration for three pions.

For $\tau$-decay $k \ll M$ and $q \ll M$; the behaviour $d\Gamma \propto k^{2l}q^{2L}$ then implies that a flat distribution can only be associated with a $0^-$ configuration. For $l > 0, L > 0$ depopulation would occur at the points $A$ and $B$ in Fig. 7.6 ($\mathbf{q} = 0$ corresponds to point $A$, $\mathbf{k} = 0$ to point $B$). An even population for the Dalitz plot is found to a good approximation in $K^+$ decay, thus indicating a spin-parity configuration $0^-$ for the $3\pi$ system. This result shows the spin of the kaon to be zero since angular momentum must be conserved; it gives no information on the parity of the kaon since this is a weak interaction and parity conservation is violated. The parity of the kaon has been shown to be negative from a study of the properties of the kaon in strong interactions.

Now consider $\omega$-decay

$$\omega \to \pi^+\pi^-\pi^0$$

This is a strong interaction and so both parity and isospin must be conserved. Now the $\omega$ meson only exists in the change zero state and so has isospin zero; since the $\pi^0$ meson has isospin $I = 1, I_3 = 0$, then the $\pi^+ \pi^-$ combination must also have $I = 1, I_3 = 0$ so that $I_\omega = 0$. Use of the Clebsch–Gordan coefficients [see § 7.2(a)] then gives the following state for $\pi^+\pi^-$:

$$|I = 1, I_3 = 0\rangle = \frac{1}{\sqrt{2}}(|\pi^+\pi^-\rangle - |\pi^-\pi^+\rangle) \tag{7.1.26}$$

and it is evident that this system has odd symmetry under the exchange $\pi^+ \leftrightarrow \pi^-$.‡ Since we are dealing with pions the overall wave function for the $\pi^+ \pi^-$ pair must have even symmetry to satisfy Bose–Einstein statistics. Now for spinless particles exchange is physically equivalent to the reflection of coordinates (the parity operation) and so the $(-1)^L$ relationship of equation (6.1.5) limits us to odd values of $L$, i.e. odd powers of $q$ in order to maintain overall symmetry. The experimental data for $\omega$-decay gives a depopulation around the edges of the Dalitz plot (Fig. 7.8). The simplest term which will represent this is $\mathbf{k} \times \mathbf{q}$, since

‡ Although we have chosen to discuss the $\pi^+\pi^-$ pair here, the same considerations apply to any pair of pions in the decay of an isospin zero system. Thus the isospin function must be antisymmetric under the exchange of any pair of pions. The same consideration must then apply to the space part of the wave function.

when **k** and **q** are collinear the kinematic point is at the boundary of the plot (for a given **k**, $T_2$ has a maximum value and $T_3$ a minimum or vice versa according to Fig. 7.7). The expression $\mathbf{k} \times \mathbf{q}$ is an axial vector and since the three pions have intrinsic parity $(-1)^3$ then $j^P$ for the $\omega$ meson is $1^-$. An alternative way of reaching this conclusion is to construct a table

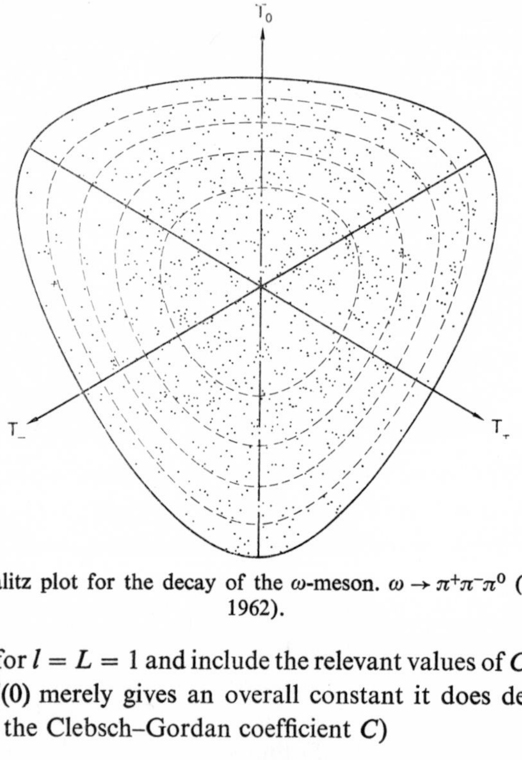


FIG. 7.8. Dalitz plot for the decay of the $\omega$-meson. $\omega \to \pi^+\pi^-\pi^0$ (Alff *et al.*, 1962).

as in (7.1.25) for $l = L = 1$ and include the relevant values of $CY_l^0(0)Y_L^m(\theta)$ (although $Y_l^0(0)$ merely gives an overall constant it does determine the behaviour of the Clebsch–Gordan coefficient $C$)

| $j^P$ for $\omega$ | $l$ | $L$ | $CY_l^0(0)\, Y_L^m(\theta)$ | $\|f\rangle$ | |
|---|---|---|---|---|---|
| $0^-$ | 1 | 1 | $Y_1^0 \equiv \cos\theta$ | $\mathbf{k}\cdot\mathbf{q}$ | (7.1.27) |
| $1^-$ | 1 | 1 | $Y_1^1 \equiv \sin\theta$ | $\mathbf{k}\times\mathbf{q}$. | |

Notice that since the isospin function is completely antisymmetric the expressions for $|f\rangle$ must be fully antisymmetrised. We have included them in this form in Table 7.1 below.

At first sight the technique described above may not seem to be satisfactory since the approach made was essentially non-relativistic. This is not so since we have been working in the rest frame of the particle and as we have shown in Chapter 3 spin has the conventional meaning in this frame. Consider the Lorentz invariant form of the matrix element for $\omega$-decay. Isospin considerations [see the remarks prior to (7.1.26)] force us to choose a completely antisymmetric final state for the spatial terms. Recalling the techniques discussed in § 4.2 a suitable matrix element for a $j^P = 1^-$ system is

$$T_{fi} = \gamma\, \varepsilon_{\mu\nu\lambda\sigma}\, e_\mu k_{1\nu} k_{2\lambda} k_{3\sigma} \tag{7.1.28}$$

[compare (4.2.2)], where $\gamma$ is a Lorentz invariant function of the dynamical variables (coupling strengths, etc.), $\varepsilon$ is the completely antisymmetric tensor [compare (3.4.1)], $e_\mu$ the $\omega$ spin and $k_1$, $k_2$ and $k_3$ the four momenta of the three pions. In the $\omega$ rest frame $e_4 = 0$ and $T_{fi}$ reduces to

$$T_{fi} = i\gamma\, \mathbf{e} \cdot (\mathbf{k}_1 \times \mathbf{k}_2\, E_3 + \mathbf{k}_2 \times \mathbf{k}_3\, E_1 + \mathbf{k}_3 \times \mathbf{k}_1\, E_2)$$

where $E_1$, $E_2$, $E_3$ are the pion energies. In the rest frame we also have

$$M_\omega = E_1 + E_2 + E_3 \qquad \mathbf{k}_3 = -\,\mathbf{k}_1 - \mathbf{k}_2$$

hence

$$T_{fi} = 3i\gamma M_\omega \mathbf{e} \cdot \mathbf{k}_1 \times \mathbf{k}_2 = 3i\gamma M_\omega \mathbf{e} \cdot \mathbf{k} \times \mathbf{q} \tag{7.1.29}$$

since components of vectors perpendicular to Lorentz transformations remain invariant. Thus we have recovered our dynamical variable of (7.1.27).

An extensive analysis of the decay of resonances to many body final states has been made by Zemach (1964, 1965). In this work the fact that a symmetrised three tensor of rank $j$ (integral) behaves under rotations like a particle of spin $j$ in its rest frame has been extensively exploited. The appropriate expressions are then non-relativistic in form and equivalent to those which arise in our tabulations (7.1.25) and (7.1.27). The technique is discussed in Appendix A7. The values for $|f\rangle$ for an isospin zero system with $j = 0, 1$ are given in Table 7.1. The boost prescriptions (Appendix A5) can then be used to provide Lorentz invariant forms if necessary.

TABLE 7.1.

| $j^P$ | $L$ | $l$ | $\|f\rangle$ Leading terms → Antisymmetrical form | Zeros on Dalitz plot |
|---|---|---|---|---|
| $0^-$ | 1 | 1 | $\mathbf{k}\cdot\mathbf{q} \rightarrow (E_1 - E_2)(E_2 - E_3)(E_3 - E_1)$ | medians |
| $1^+$ | 1 | 0 | $\mathbf{q} \rightarrow E_3(\mathbf{k}_1 - \mathbf{k}_2) + E_1(\mathbf{k}_2 - \mathbf{k}_3) + E_2(\mathbf{k}_3 - \mathbf{k}_1)$ | $q = 0$ and centre |
| $1^-$ | 1 | 1 | $\mathbf{k}\times\mathbf{q} \rightarrow \mathbf{k}_1 \times \mathbf{k}_2 + \mathbf{k}_2 \times \mathbf{k}_3 + \mathbf{k}_3 \times \mathbf{k}_1$ | boundary |

## 7.2. Internal Quantum Numbers

The spin and parity of a particle are concerned with its spatial properties—spin with rotations and parity with reflections. In addition the hadrons have certain internal properties which help to determine the way in which they interact and which are unrelated to space-time.

We have already encountered one such property—that of charge conjugation. For example, the fact that the $\varrho$-meson decays to two pions and has spin $j = 1$ implies (6.1.8)

$$C\,|\varrho^0\rangle = C\,|\pi^+\pi^-\rangle = -|\pi^+\pi^-\rangle = -|\varrho^0\rangle.$$

### 7.2(a). Isospin

The concept of isospin was first suggested by Heisenberg (1932) soon after the discovery of the neutron. It was found that the binding energy of nuclei was substantially the same if a neutron was substituted for a proton, e.g.

$$\text{binding energy } H^3 = 8{\cdot}49 \text{ MeV}, \quad He^3 = 7{\cdot}73 \text{ MeV}.$$

This result led to the concept of the charge independence of nuclear forces. Heisenberg proposed that the states of proton and neutron were really labels describing different states of a single system—the nucleon. He further suggested that the labels $p$ and $n$ merely describe "spin-up" and "spin-down" states of a nucleon in a special space—the isospin space.

States with two orientations can be regarded as spin $\frac{1}{2}$ systems and if we pursue the spin analogy a little further we can take over the algebra of the Pauli spin operators (3.5.5)

ordinary spin → isospin

operator $\boldsymbol{\sigma} \to \boldsymbol{\tau}$

$$|p\rangle = \begin{pmatrix} 1 \\ 0 \end{pmatrix} \qquad |n\rangle = \begin{pmatrix} 0 \\ 1 \end{pmatrix}$$

$$\tau_{\pm} = \frac{1}{2}(\tau_1 \pm i\tau_2) \qquad \tau_3 = \begin{pmatrix} 1 & 0 \\ 0 & -1 \end{pmatrix}$$

$$\tau_3\,|p\rangle = |p\rangle \qquad \tau_3\,|n\rangle = -|n\rangle$$

$$\tau_-\,|p\rangle = |n\rangle \qquad \tau_+\,|n\rangle = |p\rangle. \tag{7.2.1}$$

Now consider pions—the pion has three charged states $-1, 0, 1$ and so can be regarded as a system with isospin 1. Using the angular momentum analogy the three charged states of the pion can be expressed as

$$\pi^+ \equiv Y_1^1 = -\sqrt{\frac{3}{8\pi}}\sin\theta e^{i\phi} \equiv -\frac{1}{\sqrt{2}}\frac{1}{r}(x+iy) \equiv -\frac{1}{\sqrt{2}}(\pi_1 + i\pi_2)$$

$$\pi^- \equiv Y_1^{-1} = \sqrt{\frac{3}{8\pi}}\sin\theta e^{i\phi} \equiv \frac{1}{\sqrt{2}}\frac{1}{r}(x-iy) \equiv \frac{1}{\sqrt{2}}(\pi_1 - i\pi_2)$$

$$\pi^0 \equiv Y_1^0 = \sqrt{\frac{3}{4\pi}}\cos\theta \equiv \frac{z}{r} \equiv \pi_3. \tag{7.2.2}$$

It can be seen that we have a close analogy with the components of a vector and it is customary to refer to the pion field as a vector field in isospin space.

We complete the analogy with ordinary spin by assuming that the isospin operator $\mathbf{I}^2$ remains invariant under rotations in isospin space just as $\mathbf{J}^2$ does for rotations in ordinary space and that combinations of isospin states are linked by the same Clebsch–Gordan coefficients as are angular momentum states (compare 7.1.18)

$$|i_a, i_{3a};\ \ i_b, i_{3b}\rangle = \sum_I C_{I\ \ i_a\ \ i_b}^{I_3\ i_{3a}\ i_{3b}}\,|I, I_3\rangle \tag{7.2.3}$$

The linking of $I_3$ with charge for any hadron system is defined through the Gell-Mann, Nishijima relation (1.4.3)

$$\frac{Q}{e} = I_3 + \frac{1}{2}(B + S).$$

### 7.2(b). Isospin in π-p scattering

The strongest justification for the use of isospin in strong interactions arose from the early experiments on pion–nucleon scattering. These led to the discovery of the $j^P = 3/2^+$ resonance at 1236 MeV; its isospin was determined by measurements of $\pi^{\pm}p$ elastic scattering.

We start with our standard matrix element

$$T_{fi} = \langle f | T | i \rangle \qquad f, i \equiv \pi N$$

and insert a complete set of isospin states, just as we inserted a complete set of angular momentum states in (7.1.2)

$$T_{fi} = \sum_{I', I'_3} \sum_{I, I_3} \langle f | I', I'_3 \rangle \langle I', I'_3 | T | I, I_3 \rangle \langle I, I_3 | i \rangle \qquad (7.2.4)$$

Since we are assuming charge independence, $T$ must be independent of $I_3$ because the latter is linearly related to charge through the Gell-Mann, Nishijima relation (1.4.3). Since we also assume conservation of $I$ and $I_3$ in strong interactions

$$I' = I \qquad I'_3 = I_3$$

then

$$\langle I, I_3 | T | I', I'_3 \rangle \rightarrow \langle I | T | I \rangle \rightarrow T_I.$$

The terms $\langle f | I, I_3 \rangle$ and $\langle I, I_3 | i \rangle$ in (7.2.4) are simply Clebsch–Gordan coefficients (7.2.3)

$$\langle f | I, I_3 \rangle = C_{fII_3} \equiv C_{I\ i_a\ i_b}^{I_3\ i_{3a}\ i_{3b}}$$

which can be looked up in appropriate tables, and so equation (7.2.4) becomes

$$T_{fi} = \sum_{I} C_{fII_3} T_I C_{iII_3}. \qquad (7.2.5)$$

As an example consider $\pi^{\pm}p$ scattering. In this case two isospin states are possible, $\frac{1}{2}$ and $\frac{3}{2}$, since the pion and proton have isospin 1 and $\frac{1}{2}$

respectively. For $\pi^+p$ scattering only one state is possible $I = \frac{3}{2}$, since $I_3 = \frac{3}{2}$ for $\pi^+(i_3 = 1)$ and proton $(i_3 = \frac{1}{2})$; for $\pi^-p$ both $I = \frac{3}{2}$ and $\frac{1}{2}$ are present and tables of Clebsch–Gordan coefficients must be used (Appendix A3)

$$\pi^+p \to \pi^+p \qquad T_{fi} = 1\, T_{3/2}\, 1$$

$$\pi^-p \to \pi^-p \qquad T_{fi} = \sqrt{\frac{1}{3}}\, T_{3/2} \sqrt{\frac{1}{3}} + \sqrt{\frac{2}{3}}\, T_{1/2} \sqrt{\frac{2}{3}}$$

$$= \frac{1}{3} T_{3/2} + \frac{2}{3} T_{1/2}$$

$$\pi^-p \to \pi^0 n \qquad T_{fi} = \sqrt{\frac{2}{3}}\, T_{3/2} \sqrt{\frac{1}{3}} - \sqrt{\frac{1}{3}}\, T_{1/2} \sqrt{\frac{2}{3}}$$

$$= \sqrt{\frac{2}{9}}\, T_{3/2} - \sqrt{\frac{2}{9}}\, T_{1/2}.$$

Thus if we make extreme assumptions about the different isospin channels we find

$$\sigma_{\pi^+\to\pi^+} : \sigma_{\pi^-\to\pi^-} : \sigma_{\pi^-\to\pi^0} = 1 : \frac{1}{9} : \frac{2}{9} \qquad I = \frac{3}{2}\ \text{dominance}$$

$$= 0 : \frac{4}{9} : \frac{2}{9} \qquad \frac{1}{2}$$

$$= 195:22:45\ \text{mb} \quad \text{experiment at } m_R = 1236\ \text{MeV}.$$

The experimental figures quoted above correspond to the lowest energy peaks observed in the $\pi p$ scattering cross-sections. The data clearly shows that this resonance has isospin $I = \frac{3}{2}$; in addition the angular distribution and polarisation give spin parity $j^P = \frac{3}{2}^+$. Thus the quantum numbers of the resonance are completely determined.

If we considered all possible charged states of the pion–nucleon system we would find eight possible channels for $\pi N \to \pi N$. Without the concept of isospin invariance we would have eight independent amplitudes; isospin reduces this number to two, and is thus an important unifying factor in the study of strong interactions.

### 7.2(c). *G*-parity

We remarked in § 6.1(b) that the $\pi^0$ meson had even charge parity $C\,|\pi^0\rangle = |\pi^0\rangle$ (6.1.12). It was pointed out by Lee and Yang (1956b) that the usefulness of the $C$ operation could be extended for pion systems by combining $C$ with a rotation by $\pi$ radians about the 2 axis in isospin space. The combined operation is called $G$ conjugation

$$G = e^{-i\pi I_2}\, C$$

[compare (3.3.17) for an example of a rotation]. We shall represent the pions as in (7.2.2)

$$|\pi^+\rangle \equiv \frac{-1}{\sqrt{2}}(|\pi_1\rangle + i\,|\pi_2\rangle)$$

$$|\pi^-\rangle \equiv \frac{1}{\sqrt{2}}(|\pi_1\rangle - i\,|\pi_2\rangle)$$

$$|\pi^0\rangle \equiv |\pi_3\rangle.$$

Now it is evident from the above format that charge conjugation must change the sign of $\pi_2$ in $\pi^+$ and $\pi^-$

$$C\,|\pi^\pm\rangle = \zeta_C\,|\pi^\pm\rangle = -|\pi^\pm\rangle$$

whilst a rotation by $\pi$ radians about the $I_2$ axis induces the changes $\pi_1 \to -\pi_1$, $\pi_2 \to \pi_2$, $\pi_3 \to -\pi_3$; hence

$$|\pi^+\rangle \xrightarrow{C} \frac{-1}{\sqrt{2}}(|\pi_1\rangle - i\,|\pi_2\rangle) \xrightarrow{I_2(\pi)} \frac{-1}{\sqrt{2}}(-|\pi_1\rangle - i\,|\pi_2\rangle) = -|\pi^+\rangle$$

$$|\pi^-\rangle \qquad \frac{1}{\sqrt{2}}(|\pi_1\rangle + i\,|\pi_2\rangle) \qquad \frac{1}{2}(-|\pi_1\rangle + i\,|\pi_2\rangle) = -|\pi^-\rangle$$

$$|\pi^0\rangle \qquad |\pi_3\rangle \qquad -|\pi_3\rangle = -|\pi^0\rangle.$$

Thus we find

$$G\,|\pi\rangle = -|\pi\rangle$$

for any charge, and for $n$ pions

$$G\,|n\pi\rangle = (-1)^n\,|n\pi\rangle. \tag{7.2.6}$$

The concept of $G$ parity is useful in providing an additional quantum number for resonances decaying by strong interactions into pions only, and for limiting the possible forms of interaction taking place at a vertex. It is not possible, for example, to consider a vertex involving an ingoing and outgoing pion and pion propagator when constructing Feynman amplitudes (§ 4.4) since $n = 3$; a vertex constructed from two pions and a rho meson is admissible, however ($\equiv n = 4$).

The $G$ conjugation operation is obviously limited to systems with $B = S = 0$, and can therefore apply to proton–antiproton combinations. The rule for this combination is

$$G\,|p\bar{p}\rangle = (-1)^{l+s+I}\,|p\bar{p}\rangle$$

where $l$, $s$, $I$ refer to the orbital angular momentum, spin and isospin of the $p\bar{p}$ pair.

### 7.2(d). Higher symmetries

We have seen in pion–nucleon scattering that the concepts of isospin multiplets and conservation of isospin in strong interactions can produce order amongst the otherwise uncorrelated scattering amplitudes in the various charge states. These ideas are obviously powerful tools in strong interaction physics.

The discovery of the strange particles and then the resonances has led to a search for systems of higher symmetry which can lead to some sort of order amongst otherwise terrifying accumulation of bits of miscellaneous data.

Such systems appear to exist. Some time after the discovery of the Gell-Mann, Nishijima equation (1.4.3), it was pointed out by d'Espagnat and Prentki (1956) that one could be more economical in the use of quantum numbers if a hypercharge number $Y$ was introduced

$$\frac{Q}{e} = I_3 + \frac{1}{2}(B + S) = I_3 + \frac{1}{2}Y. \tag{7.2.7}$$

Now if we plot $I_3$ against $Y$ for $j^P = 0^-$, $1^-$ mesons and $\frac{1}{2}^+$ baryons we find

$0^-$ mesons

| $\uparrow Y$ \ $I_3 \longrightarrow$ | $-1$ | | $0$ | | $+1$ |
|---|---|---|---|---|---|
| $+1$ | | $K^0$ | | $K^+$ | |
| $0$ | $\pi^-$ | | $\pi^0, \eta, X^0$ | | $\pi^+$ |
| $-1$ | | $K^-$ | | $\bar{K}^0$ | |

$1^-$ mesons

| $\uparrow Y$ \ $I_3 \longrightarrow$ | $-1$ | | $0$ | | $+1$ |
|---|---|---|---|---|---|
| $+1$ | | $K^{*0}$ | | $K^{*+}$ | |
| $0$ | $\varrho^-$ | | $\varrho^0, \omega, \phi$ | | $\varrho^+$ |
| $-1$ | | $K^{*-}$ | | $\bar{K}^{*0}$ | |

$\frac{1}{2}^+$ baryons

| $\uparrow Y$ \ $I_3 \longrightarrow$ | $-1$ | | $0$ | | $+1$ |
|---|---|---|---|---|---|
| $+1$ | | $n$ | | $p$ | |
| $0$ | $\Sigma^-$ | | $\Sigma^0, \Lambda$ | | $\Sigma^+$ |
| $-1$ | | $\Xi^-$ | | $\Xi^0$ | |

A remarkable correlation can therefore be seen between the different multiplets. Now if these patterns represent some higher-order symmetry than isospin one might have expected the masses of the particles within the multiplet to be the same. Obviously they are not and the differences must be caused by strong interactions since they are far too large for deviations arising from electromagnetic effects—these one might expect to be of order 1/137 (Fig. 7.9).

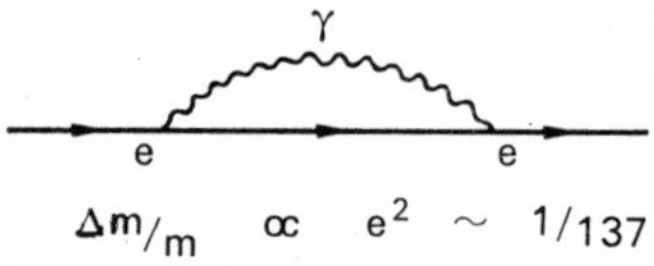


FIG. 7.9.

Let us therefore represent the mass of a particle $a$ within a multiplet as

$$M_a = M_m + \langle a| H_s |a\rangle + \langle a| H_{em} |a\rangle \tag{7.2.8}$$

where $M_m$ represents the mass of the multiplet if $H_s = H_{em} = 0$ and the latter two terms are the Hamiltonians which give rise to mass deviations by internal strong and electromagnetic interactions respectively. We shall first set $H_{em} = 0$, i.e. ignore the mass differences within isospin submultiplets and assume that $H_s$ is proportional to the relevant operators which parameterise the multiplet up to second powers

$$H_s = \alpha \mathbf{I}^2 + \beta Y + \gamma Y^2.$$

$I_3$ does not appear in this equation since it is linked with electric charge $Q$ and therefore the electromagnetic interaction. The lowest order in which isospin can occur is $\mathbf{I}^2$ and therefore it appears logical that $Y^2$ should appear. The assumption that the multiplets are contained in a higher symmetry group called $SU_3$ [§ 7.2(f)] yields a relationship between $\alpha$, $\beta$ and $\gamma$, and allows equation (7.2.8) to be cast in the form

$$M_a = M_m\{1 + \delta Y + \varepsilon[I(I+1) - \tfrac{1}{4} Y^2]\}. \qquad (7.2.9)$$

This equation is known as the Gell-Mann, Okubo mass formula (Gell-Mann, 1962; Okubo, 1962). A simple derivation is given on p. 623 of *P.E.P.*

The mass formula leads to the following relation for the baryons in the octet

$$3M_\Lambda + M_\Sigma = 2M_N + 2M_\Xi.$$

It is satisfied to better than one per cent accuracy. In the case of the mesons a more complicated situation exists. Since $Y \rightarrow -Y$ is equivalent to charge conjugation for mesons and we require invariance under charge conjugation in strong interactions then $\delta = 0$ in the mass formula. For reasons which have never been properly explained the formula works better for $M^2$ than $M$; the equation then reduces to

$$3M_0^2 + M_1^2 = 4M_{\frac{1}{2}}^2$$

where the subscripts refer to isospin values. The interpretation of this formula is bound up with the problem of mixing of isospin zero states and so discussion will be postponed until § 7.2(g).

If we go to the other extreme in equation (7.2.8) and set $H_s = 0$, then derivations from $M_m$ can only depend on the electric charge $Q$. We then expect relations of the following type from an inspection of the multiplet patterns for $\frac{1}{2}^+$ baryons

$$M_p - M_m = M_{\Sigma^+} - M_m$$

$$M_n - M_m = M_{\Xi^0} - M_m$$

$$M_{\Sigma^-} - M_m = M_{\Xi^-} - M_m$$

Hence if we arrange the masses to cancel the strong interaction effects, then

$$M_{\Xi^-} - M_{\Xi^0} = M_{\Sigma^-} - M_{\Sigma^+} + M_p - M_n$$

and the data on particle masses yield 6·5 ± 0·7 and 6·6 ± 0·2 MeV for left- and right-hand sides respectively. This relation and many others on electromagnetic properties within the $SU_3$ scheme was first given by Coleman and Glashow (1961).

A further basic pattern exists for the baryon multiplets. It is a decuplet (10 members) and is shown below for baryons with spin-parity $\frac{3}{2}^+$. An inspection of this multiplet gives the empirical relation $Y = 2(I - 1)$ and so the mass formula (7.2.9) reduces to

$$M_a = M_m(1 + \sigma Y).$$

An early success of the higher symmetry concept was the prediction of the $\Omega^-$ hyperon, and its mass from the above relation, in order to com-

| $Y$ ↑ \ $\longrightarrow I_3$ | $-\frac{3}{2}$ | $-1$ | $-\frac{1}{2}$ | $0$ | $\frac{1}{2}$ | $1$ | $\frac{3}{2}$ | Mass | Difference |
|---|---|---|---|---|---|---|---|---|---|
| 1 | $N^{*-}$ | | $N^{*0}$ | | $N^{*+}$ | | $N^{*++}$ | 1237 | |
| | | | | | | | | | 148 MeV |
| 0 | | $\Sigma^{*-}$ | | $\Sigma^{*0}$ | | $\Sigma^{*+}$ | | 1385 | |
| | | | | | | | | | 148 |
| −1 | | | $\Xi^{*-}$ | | $\Xi^{*0}$ | | | 1533 | |
| | | | | | | | | | 148 |
| $Y$ −2 | | | | $\Omega^-$ | | | | 1681 ← | (expected) |

plete the membership of this multiplet. The $\Omega^-$ hyperon was later identified in the reaction

$$K^-p \rightarrow \Omega^-K^+K^0$$

and its mass was measured to be 1686 ± 12 MeV (Barnes *et al.*, 1964). Since the $\Omega^-$ has hypercharge $Y = -2$, then its strangeness is $S = -3$ from (7.2.7).

### 7.2(e). Quarks

The problem of explaining why hadron systems prefer to group in multiplets of 9, 8 and 10 has been greatly helped by the concept of quarks.

In physics one likes to have as few basic postulates or building blocks as possible, and elementary particle physics has been helped by the knowledge that in nuclear structure physics all the possible nuclei are constructed from two systems—the neutron and proton.

The first attempt to reduce the number of "elementary" particles was made by Fermi and Yang (1949) who pointed out that the formal properties of the pion could be duplicated by a bound nucleon–antinucleon system in a singlet $s$ state

$$|\pi^+\rangle \equiv |\bar{n}p\rangle \quad j^P = 0^- \quad \text{for } {}^1S_0 \text{ state of } \bar{n} \text{ and } p \quad [\S\ 6.1(a)]$$

$$I_3 = \tfrac{1}{2} + \tfrac{1}{2} = 1$$

$$C|\pi^+\rangle = C|\bar{n}p\rangle = \zeta_C|n\bar{p}\rangle = \zeta_C|\pi^-\rangle$$

similarly

$$|\pi^-\rangle \equiv |\bar{p}n\rangle \qquad I_3 = -1$$

$$|\pi^0\rangle \equiv \frac{1}{\sqrt{2}}(|\bar{p}p\rangle - |\bar{n}n\rangle) \quad I_3 = 0$$

and we have left over an isotopic singlet

$$|\eta\rangle = \frac{1}{\sqrt{2}}(|\bar{p}p\rangle + |\bar{n}n\rangle) \quad I = 0, I_3 = 0.$$

The factor $\sqrt{2}$ appears for purposes of normalisation.

At first the ± signs may look wrong in the above equations. But it must be remembered that the pion behaves like a vector in isospin space and an isospin vector for nucleon–antinucleon systems can be constructed with the aid of the isospin operator $\boldsymbol{\tau}$

$$\boldsymbol{\pi} = \langle\bar{N}|\boldsymbol{\tau}|N\rangle \qquad (7.2.10)$$

then from (7.2.1) and (7.2.2)

$$
\begin{aligned}
\pi^+ &\equiv \pi_1 + i\pi_2 \equiv \langle \bar{N}|\, \tau_1 + i\tau_2 \,|N\rangle \equiv \langle \bar{N}|\, \tau_+ \,|N\rangle = \bar{n}p \\
\pi^- &\equiv \pi_1 - i\pi_2 \equiv \langle \bar{N}\, |\tau_1 - i\tau_2 \,|N\rangle \equiv \langle \bar{N}|\, \tau_- \,|N\rangle = \bar{p}n \qquad (7.2.11) \\
\pi^0 &\equiv \pi_3 \qquad\quad \equiv \langle \bar{N}|\, \tau_3 \,|N\rangle \qquad\qquad = \bar{p}p - \bar{n}n.
\end{aligned}
$$

Similarly the singlet can be regarded as

$$\eta \equiv \langle \bar{N}|\, \hat{1} \,|N\rangle = \bar{p}p + \bar{n}n.$$

Following the discovery of the strange particles a further particle had to be added to the basic blocks but none of the schemes worked satisfactorily until Gell-Mann (1964) and Zweig (1964) independently realised that self-consistency could be achieved by giving the basic blocks a baryon number B/3. These particles were called quarks by Gell-Mann and aces by Zweig and are given the labels $p$, $n$ and $\Lambda$ to conform to their isospin and strangeness properties. They have the following quantum numbers:

| | $I_3$ | $I$ | $S$ | | $Q/e$ | $Y$ |
|---|---|---|---|---|---|---|
| $Q_p$ | $\frac{1}{2}$ | $\frac{1}{2}$ | 0 | Common properties | $\frac{2}{3}$ | $\frac{1}{3}$ |
| $Q_n$ | $-\frac{1}{2}$ | | 0 | spin $= \frac{1}{2}$ | $-\frac{1}{3}$ | $\frac{1}{3}$ |
| $Q_\Lambda$ | 0 | 0 | $-1$ | $B = \frac{1}{3} \rightarrow$ | $-\frac{1}{3}$ | $-\frac{2}{3}$ |

(7.2.12)

We can then construct mesons as

$$|M\rangle \sim |Q\bar{Q}\rangle$$

and baryons as

$$|B\rangle \sim |QQQ\rangle.$$

Consider the model in a little more detail. If we consider the meson as a $Q\bar{Q}$ in a state of relative orbital angular momentum $L$ we can asign the following quantum numbers [§ 6.1 (a) and 6.1 (b)]

$$\mathbf{j} = \mathbf{L} + \mathbf{s} \qquad P = (-1)^{L+1} \qquad C = (-1)^{L+s}$$

hence for the two lowest-lying states we have

$$
\begin{aligned}
L = 0 \quad s &= 0 \quad \downarrow\uparrow \quad j^P = 0^- \quad C = +1 \\
s &= 1 \quad \uparrow\uparrow \qquad\; = 1^- \qquad\; = -1
\end{aligned}
$$

similarly for $QQQ$ triplet the lowest-lying states are

$$L = 0 \quad s = \tfrac{1}{2} \quad \uparrow\uparrow\downarrow \quad j^P = \tfrac{1}{2}^+$$

$$s = \tfrac{3}{2} \quad \uparrow\uparrow\uparrow \quad\quad = \tfrac{3}{2}^+$$

These four spin-parity assignments account for all particles listed in § 7.2(d). When $L \neq 0$ further systems can be constructed; these systems can be identified with known states and empirically one finds that as $L$ increases the mass increases.

We now return to the problem of why the multiplets contain 9, 8 and 10 particles. We shall use an argument based on patterns. Consider the coupling of two isospin one systems. We can represent one system as

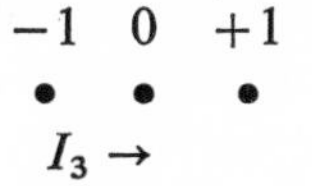

$$
\begin{array}{ccc}
-1 & 0 & +1 \qquad i = 1. \\
\bullet & \bullet & \bullet \\
I_3 \rightarrow & &
\end{array}
$$

Now we lay another $i = 1$ system on each of these three points and find

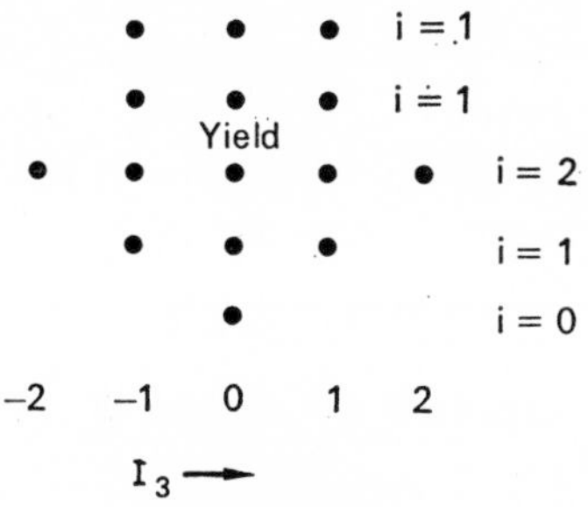

FIG. 7.10.

thus we generate a population of nine points which break down into multiplets of $5(i = 2)$, $3(i = 1)$ and $1(i = 0)$. We may represent this as

$$\underset{\sim}{3} \otimes \underset{\sim}{3} = \underset{\sim}{5} + \underset{\sim}{3} + \underset{\sim}{1}.$$

Now consider $Q\bar{Q}$ systems, we have two basic triplets (Fig. 7.11); if we place the triangle on each of the points of the other we find the pattern shown in Fig. 7.12. The centre point is confusing; the $\Lambda\bar{\Lambda}$ is obviously an isotopic singlet and we also have $p\bar{p}$, $n\bar{n}$ which are formally indis-

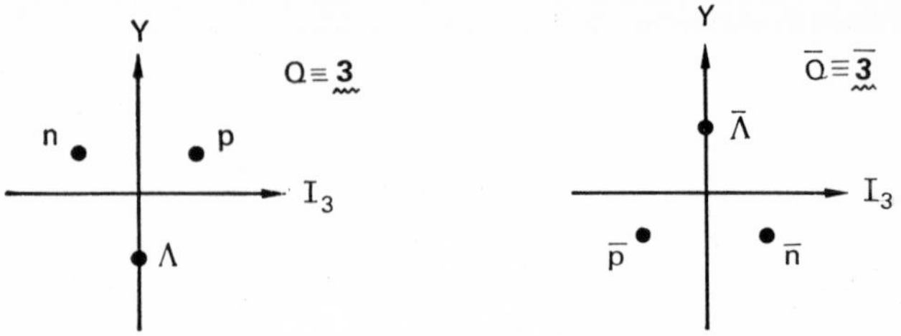


FIG. 7.11.

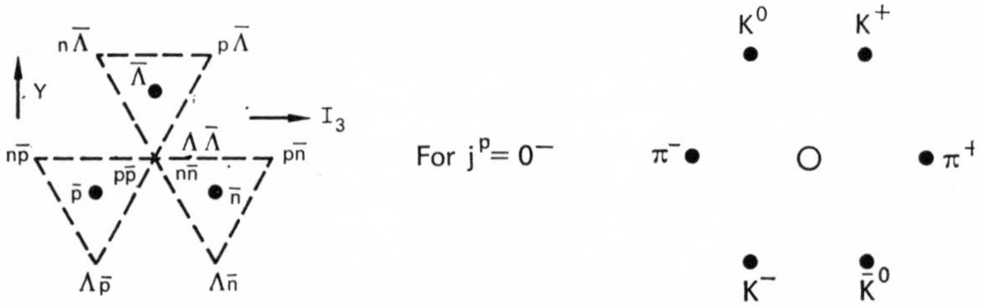


FIG. 7.12.

tinguishable and comprise a system of mixed isospin states one of which must be a $\pi^0$ and the other an isotopic singlet [compare (7.2.11)]. Ignoring the problem of identification for the moment [we shall return to it in § 7.2(g)], we have effectively evaluated

$$\underset{\sim}{\mathbf{3}} \otimes \underset{\sim}{\bar{\mathbf{3}}} = \underset{\sim}{\mathbf{1}} + \underset{\sim}{\mathbf{8}}. \tag{7.2.13}$$

Now consider what happens if we examine $\underset{\sim}{\mathbf{3}} \otimes \underset{\sim}{\mathbf{3}}$. We find the result shown in Fig. 7.13 and so

$$\underset{\sim}{\mathbf{3}} \otimes \underset{\sim}{\mathbf{3}} = \underset{\sim}{\mathbf{6}} + \underset{\sim}{\bar{\mathbf{3}}}.$$

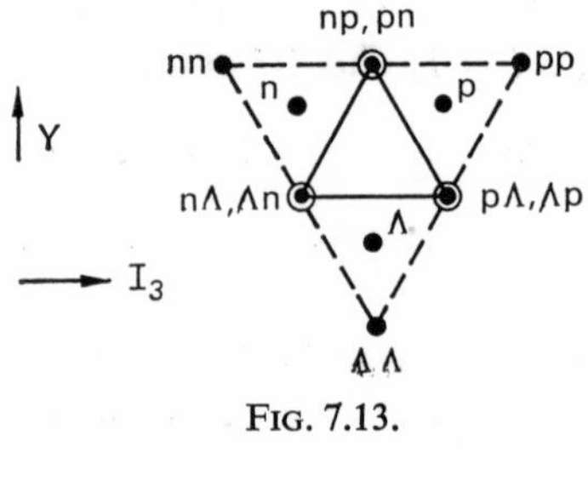


FIG. 7.13.

The whole operation can be repeated giving

$$\underset{\sim}{\mathbf{3}} \otimes \underset{\sim}{\mathbf{3}} \otimes \underset{\sim}{\mathbf{3}} = (\underset{\sim}{\mathbf{6}} + \underset{\sim}{\bar{\mathbf{3}}}) \otimes \underset{\sim}{\mathbf{3}} = \underset{\sim}{\mathbf{10}} + \underset{\sim}{\mathbf{8}} + \underset{\sim}{\mathbf{8}} + \underset{\sim}{\mathbf{1}}. \qquad (7.2.14)$$

The constituents of the 10 system can be easily identified. Laying the Q = **3** diagram of Fig. 7.11 on Fig. 7.13 yields a triangle with quark content *nnn*, *ppp* and *ΛΛΛ* at the extremities. The points *nnn* and *ppp* have $I_3 = -\frac{3}{2}$ and $\frac{3}{2}$ respectively and $Y = 1$ from (7.2.12). Whilst *ΛΛΛ* has $I = 0$, $Y = -2$ (the quantum numbers of the $\Omega^-$ hyperon). Thus we have identified the extremities of the decuplet of § 7.2(d); the remaining points can be easily filled in. An examination of the properties of the pecuplet in more detail also shows that the spins must be ↑↑↑, i.e. for $L = 0$ a $j^P = \frac{3}{2}^+$ system. This conclusion is in accord with experiment.

### 7.2(f). $SU_3$

The reader may well be wondering by now if any algebra is associated with the higher symmetry schemes. The answer is yes, and it appears to be the algebra of $SU_3$—the special unitary group in three dimensions (more strictly the group of rank 2; in general $SU_n$ is a group of rank $n-1$).

Rather than consider the group theory aspect we shall sketch the problem in terms of transformations. In § 3.3 we considered rotations of the type (3.3.9)

$$|\psi'\rangle = U(\phi)\,|\psi\rangle$$

where for an infinitesimal rotation $\delta\phi$ in one plane the unitary operator $U$ was given by (3.3.10)

$$U(\delta\phi) = \hat{1} - iJ_{xy}\,\delta\phi \equiv \hat{1} - iJ_z\,\delta\phi_z$$

where $J$ is the angular momentum operator. We can generalise this operation to the three spatial axes and at the same time consider finite transformations; $U$ then becomes

$$U = e^{-iJ_i\phi_i} \qquad i = 1, 2, 3.$$

Now let us assume the existence of other "spaces" (isospin is an example we have already considered), and not necessarily restrict ourselves to $i = 1, 2, 3$. Changing our nomenclature $J \to F, \phi \to \varepsilon$ the general transformation is then of the form

$$U(\varepsilon) = e^{-iF_\alpha \varepsilon_\alpha} \qquad \alpha = 1 \to n; \quad \varepsilon_\alpha \text{ real.} \tag{7.2.15}$$

The operator $U$ is unitary

$$U^\dagger U = \hat{1} \qquad \sum_i U^\dagger_{ji} U_{ik} = \delta_{jk}$$

and thus conserves probability. Unitary matrices are important in quantum mechanics because the scalar product of two state vectors

$$\langle a \mid b \rangle = \sum_i \langle a_i \mid b_i \rangle$$

remains invariant under unitary transformations

$$|a'\rangle = U\,|a\rangle \qquad a'_i = U_{ij} a_j$$

$$\begin{aligned} \langle a \mid b \rangle \xrightarrow{U} \langle a' \mid b' \rangle = \langle aU \mid Ub \rangle &= \sum_{ijk} (U_{ij} a_j)^* \, U_{ik} b_k \\ &= \sum_{ijk} U^\dagger_{ji} U_{ik} a^*_j b_k \\ &= \sum_{jk} \delta_{jk} \, a^*_j b_k \\ &= \langle a \mid b \rangle \end{aligned} \tag{7.2.16}$$

The transformation $U(\varepsilon)$ can be represented by a set of $n \times n$ matrices $U_n(\varepsilon)$. The dimension $n$ is determined by the dimensionality of the smallest system which can be associated with the transformation. The unitarity condition $U^\dagger U = \hat{1}$ implies that the determinant $|\det U| = 1$, and so

$$\det U = e^{i\phi}.$$

Thus the unitary matrix $U_n(\varepsilon)$ can be written as the product of a phase factor and the special unitary matrix $SU_n(\varepsilon)$ with determinant $+1$.

$$U_n(\varepsilon) = e^{i\phi} SU_n(\varepsilon) \tag{7.2.17}$$

It is customary to choose $\phi = 0$, i.e. $\det U = 1$. Unitary matrices of dimension $n$ can be represented by $n^2$ real parameters. The unimodularity condition det $U = 1$ reduces this requirement to $n^2 - 1$, hence

$$SU_n(\varepsilon) = e^{-i\varepsilon_\alpha F_\alpha} \qquad \alpha = 1 \to n^2 - 1.$$

For angular momentum and isospin $n = 2$ and the algebra is $SU_2$.‡ The transformation can be parameterised in terms of the three Pauli matrices:

$$F_\alpha = \underset{\text{angular momentum}}{\tfrac{1}{2}\sigma_\alpha} = \underset{\text{isospin}}{\tfrac{1}{2}\tau_\alpha} \qquad \alpha = 1, 2, 3. \tag{7.2.18}$$

For $SU_3$, eight matrices (generators) are required

$$F_\alpha = \tfrac{1}{2}\lambda_\alpha \qquad \alpha = 1 \to 8 \tag{7.2.19}$$

where the $\frac{1}{2}$ was introduced by Gell-Mann (1962) in analogy with $SU_2$. We shall not use the properties of $F_\alpha$ here, the interested reader may find them listed in p. 622 of *P.E.P.*

From these basic systems one can build up the higher operators and multiplets—the irreducible representations of the group. In the theory of angular momentum ($SU_2$), for example, these are the operators $J$ and states $|jm\rangle$ of dimensionality $2j + 1$. The dimensionality of the representations is obtained by determining the irreducible representations, i.e. if we form products of our basic $SU_n$ systems and apply the unitary transformations to them then the irreducible representations are those parts of the product which transform only into themselves.

An example of this behaviour was demonstrated in equation (7.2.16)—the scalar product $\langle a \mid b\rangle$ transforms into itself. Thus this product behaves like a singlet, and so for any system $SU_n$ the product of two states $a_i^* b_j (i, j = 1 \to n)$ breaks into a singlet and a system of dimensionality $n^2 - 1$. Now the antiparticle of a particle $a$ behaves like $a^*$ [compare (7.2.2) and § 3.6]. Thus for $SU_2$ and the basic doublets $\overline{N}$, $N$ we expect a singlet and a triplet for $a_i^* b_j$—as we have seen already in § 7.2(e); for $SU_3$ we expect a singlet and an octet [compare (7.2.13)]. Products of the type $a_i a_j$ and $a_i a_j a_k$ can also be reduced and the latter yields systems with dimensionality $10 + 8 + 8 + 1$ for $SU_3$. Thus we can see that the algebra corresponds to the patterns of the previous section.

‡ Mathematically this group is identical with $R_3$—the rotation group in three dimensions (see *P.E.P.*, p. 616).

### 7.2(g). Mixing angles

Let us consider the centre of the quark pattern for mesons produced by $\bar{Q}Q$ in § 7.2(e). At that point we have a combination of $\bar{\Lambda}\Lambda$, $\bar{p}p$ and $\bar{n}n$. Our previous discussion [§ 7.2(e)] has shown that the $I_3 = 0$ component of the triplet isospin state ($\pi^0$ for the $0^-$ nonet, $\varrho^0$ for the $1^-$) has the quark content $\frac{1}{\sqrt{2}}(\bar{p}p - \bar{n}n)$. The $\bar{\Lambda}\Lambda$ has isospin zero and must therefore be associated with isospin zero systems. An $SU_3$ singlet [§ 7.2(f)] can be formed from $\frac{1}{\sqrt{3}}$ $(\bar{n}n + \bar{p}p + \bar{\Lambda}\Lambda)$, where the $\sqrt{3}$ appears for normalization requirements. The remaining state required to complete the octet must then be orthogonal to the previous two and be properly normalised. The $SU_3$ content of the two isospin zero states must therefore be

$$
\begin{aligned}
|1\rangle &= \frac{1}{\sqrt{3}}(|\bar{p}p\rangle + |\bar{n}n\rangle + |\bar{\Lambda}\Lambda\rangle) \\
|8\rangle &= \frac{1}{\sqrt{6}}(|\bar{p}p\rangle + |\bar{n}n\rangle - 2\,|\bar{\Lambda}\Lambda\rangle)
\end{aligned}
\qquad (7.2.20)
$$

where $|1\rangle$ and $|8\rangle$ refer to membership of the $SU_3$ singlet and octet respectively.

In § 7.2(d) we obtained the mass relation for mesons

$$3M_0^2 + M_1^2 = 4M_{\frac{1}{2}}^2$$

where the subscripts referred to the isospin values. Now the mass of the $I = 0$ system must relate to the state $|8\rangle$ and so we can rewrite this equation as

$$3M_{08}^2 + M_1^2 = 4M_{\frac{1}{2}}^2. \qquad (7.2.21)$$

Now consider the $1^-$ nonet; here we have mesons with masses

$$
\begin{aligned}
m_\varrho &= 765 \text{ MeV} \qquad & m_{K^*} &= 891 \text{ MeV} \\
m_\omega &= 783 \text{ MeV} \qquad & m_\phi &= 1019 \text{ MeV}
\end{aligned}
$$

and if we insert these values in the relation

$$3M_{08}^2 + M_\varrho^2 = 4M_{K^*}^2 \qquad (7.2.22)$$

we obtain $M_{08} = 929$ MeV, i.e. midway between $M_\omega$ and $M_\phi$.

This is not surprising. The mesons $\omega$ and $\phi$ have the same quantum numbers

$$j^P = 1^- \quad I = 0 \quad C = -1 \quad G = -1$$

and so we cannot make an *a priori* decision on which meson belongs to the octet and which to the singlet. Since the $|\omega\rangle$ and the $|\phi\rangle$ states are orthogonal, and also the $|1\rangle$ and $|8\rangle$, we can parameterise them by a rotation of axes and a mixing angle $\theta$ (Fig. 7.14)

$$\begin{aligned} |8\rangle &= |\phi\rangle \cos\theta + |\omega\rangle \sin\theta \\ |1\rangle &= -|\phi\rangle \sin\theta + |\omega\rangle \cos\theta \end{aligned} \tag{7.2.23}$$

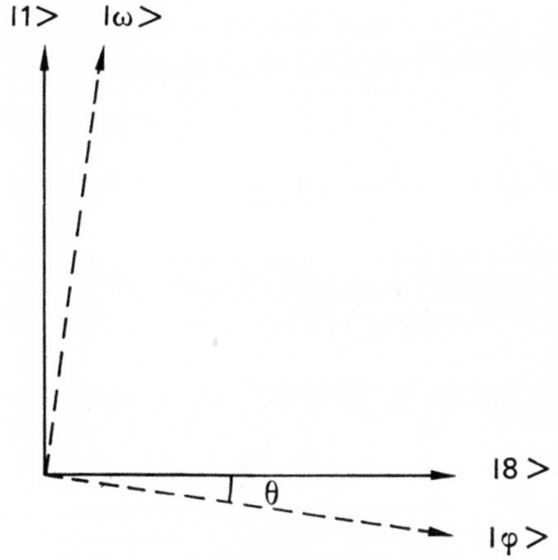


FIG. 7.14.

Thus the physical $|\omega\rangle$ and $|\phi\rangle$ represent a mixture of octet and singlet states

$$\begin{aligned} |\phi\rangle &= |8\rangle \cos\theta - |1\rangle \sin\theta \\ |\omega\rangle &= |8\rangle \sin\theta + |1\rangle \cos\theta. \end{aligned} \tag{7.2.24}$$

Defining‡

$$M_{08}^2 = \langle 8|\, H^2\, |8\rangle$$

‡ We are slightly oversimplifying matters here in order to bring out quickly the main points in the argument.

where $H$ is the total Hamiltonian, then from (7.2.23)

$$M_{08}^2 = M_\phi^2 \cos^2\theta + M_\omega^2 \sin^2\theta$$

hence equation (7.2.22) becomes

$$4M_{K^*}^2 - M_\varrho^2 = 3(M_\phi^2 \cos^2\theta + M_\omega^2 \sin^2\theta).$$

Insertion of the physical values for $M_\phi$ and $M_\omega$ then yields $\theta = 40°$.

The parameter $\theta$ can be obtained from a number of situations, for example consider the decay of a vector meson to an electron pair (Fig. 7.15). Since the photon couples to the electric current which behaves

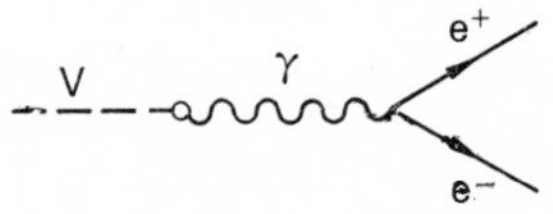


FIG. 7.15.

like a member of an octet (consider, for example, the role of the charge in the Gell-Mann, Nishijima relation), we therefore have from (7.2.24)

$$T_{fi}(\omega \to e^+e^-) = \langle e^+e^-| T |\omega\rangle = \sin\theta\langle e^+e^-| T |8\rangle$$

$$T_{fi}(\phi \to e^+e^-) = \langle e^+e^-| T |\phi\rangle = \cos\theta\langle e^+e^-| T |8\rangle$$

and if we ignore mass differences between $\phi$ and $\omega$

$$\frac{\Gamma(\omega \to e^+e^-)}{\Gamma(\phi \to e^+e^-)} = \tan^2\theta.$$

Colliding beam experiments using incident $e^+e^-$ and the direct observation of the branching ratios of $\omega$ and $\phi$ mesons to $e^+e^-$ lead to $\theta = 32 \pm 5°$ (Ting, 1968; Augustin *et al.*, 1969b).

These angles are close to the "ideal" mixing angle $\theta = 33°$, which has the following values

$$\cos\theta = \sqrt{\frac{2}{3}} \qquad \sin\theta = \sqrt{\frac{1}{3}}$$

then from (7.2.20) and (7.2.24)

$$|\phi\rangle = -|\bar{\Lambda}\Lambda\rangle$$
$$|\omega\rangle = \frac{1}{\sqrt{2}}(|\bar{p}p\rangle + |\bar{n}n\rangle). \tag{7.2.25}$$

The fact that the $\phi$ is virtually pure $\bar{\Lambda}\Lambda$ could explain its strong preference for decaying into $K\bar{K}$ pairs, which then allows the preservation of the strange quarks

$$\phi \rightarrow K\bar{K},$$
$$(\bar{\Lambda}\Lambda) \rightarrow (\bar{\Lambda}N)(\Lambda\bar{N}).$$

### 7.2(h). Summing up

The successes and failures of the $SU_3$-quark model are often confusing since one is often not sure whether they arise from the $SU_3$ or quark parts—frequently the same result can be obtained by more than one approach.

Consider some of the results

1. $SU_3$ on its own does not give the experimentally observed multiplets. $SU_3$ yields

| | | | | | | | | | |
|---|---|---|---|---|---|---|---|---|---|
| multiplet | 1, | 3, | 6, | 8, | 10 | – | – | 27 | – |
| observed | ✓ | × | × | ✓ | ✓ | | | × | |

The 27 multiplet contains exotic systems like mesons with $S = 0, I = 2$. These have not been observed. The $|Q\bar{Q}\rangle$ systems are limited to multiplicities of 1 and 8, and $|QQQ\rangle$ to 1, 8 and 10 when combined with $SU_3$. On the other hand, no convincing experimental evidence for the existence of quarks has been obtained.

2. $SU_3$ on its own gives the appropriate algebra for the invariance properties and the Clebsch–Gordan coefficients (de Swart, 1963)

$$|\alpha;\beta\rangle = \sum_{\gamma} C^{\gamma\alpha\beta}|\gamma\rangle \tag{7.2.26}$$

where $\alpha$, $\beta$ and $\gamma$ are $SU_3$ multiplets. The $SU_3$ coefficients $C$ provide numerical relations between members of multiplets, just as the Clebsch–Gordan coefficients for $SU_2$ did for members of the isospin multiplets [compare § 7.2(b)]. A typical result with $SU_3$ coefficients is the relation (Meshkov *et al.*, 1963)

$$T(\pi^- p \to N^{*-}\pi^+) = -\sqrt{3}\, T(\pi^- p \to \Sigma^{*-}K^+)$$

where the baryons in the final states are members of the $3/2^+$ decuplet. The $SU_3$ symmetry is badly violated by strong interactions and comparisons between experiment and theory do not yield such satisfactory results as the comparisons with isospin symmetry, where the breaking of the symmetry by electromagnetic interactions is relatively weak.

Certain relations between scattering amplitudes can be obtained by simply adding quarks and making no requirements from $SU_3$. Consider high energy elastic scattering at angle $\theta = 0$ for the reaction $ab \to ab$, and let us treat it as a problem in quark–quark scattering

$$T_{fi} = \langle ab|\, T\, |ab\rangle = \sum_{k,l} \langle Q_k^{ab}|\, T_{kl}\, |Q_l^{ab}\rangle \tag{7.2.27}$$

where $Q_k^{ab}$ represents the $k$th quark in the quark content of $ab$. Now if we assume that $T_{kl}$ equals a constant, $T$, for all $k$ and $l$, then in $\pi N$ scattering we would have a sum

$$T_{fi} = 6T$$

and in $NN$ scattering

$$T_{fi} = 9T.$$

At high energies the forward scattering amplitude is almost purely imaginary (§ 7.3 and Fig. 7.17)

$$T_{fi}(\theta = 0) \sim \text{Im}\, T_{fi}(\theta = 0)$$

and this result can in turn be related to total cross-sections $\sigma_T$ through the optical theorem (7.1.9), hence

$$\sigma_T(\pi p) = \tfrac{2}{3}\, \sigma_T(pp) \tag{7.2.28}$$

This prediction agrees quite well with experiment at high energies. Various refinements of the method are possible, but again not all the predictions agree with experiments and most can be obtained by other methods.

3. The additivity principle can be extended to other situations. The absence of baryon resonances with strangeness $S = 1$ in $K^+N$ formation experiments [§ 7.1(a)] follows naturally from this principle. Consider next the magnetic moments of the nucleons. We assume that they are made from the magnetic moments of their constituent quarks

$$\mu_N = g_p\langle S_{zp}\rangle + g_n\langle S_{zn}\rangle \tag{7.2.29}$$

where $g$ is the quark $g$ factor (we shall use subscripts $N$, $p$ and $n$ for the physical nucleon, the quark proton and quark neutron respectively). $\mathbf{S}_p$ and $\mathbf{S}_n$ refer to the total spins of the proton and neutron quarks respectively, and

$$\mathbf{S}_N = \mathbf{S}_p + \mathbf{S}_n.$$

Thus if we use the familiar manipulations of atomic physics

$$\begin{aligned}\mu_N &= \frac{1}{2}(g_p + g_n)\langle S_{zN}\rangle + \frac{1}{2}(g_p - g_n)\langle S_{zp} - S_{zn}\rangle \\ &= \frac{1}{2}(g_p + g_n)\langle S_{zN}\rangle + \frac{1}{2}(g_p - g_n)[S_p(S_p + 1) - S_n(S_n + 1)] \\ &\qquad \times \frac{\langle S_{zN}\rangle}{S_N(S_N + 1)}.\end{aligned}$$

If the nucleon wave function is assumed to be symmetric in spin and isospin, the two identical quarks in the nucleon must have the symmetric spin coupling, and so

$$\begin{aligned}S_p &= 1, \quad S_n = \tfrac{1}{2} \quad \text{for physical proton} \quad N \equiv ppn \\ &= \tfrac{1}{2}, \quad\;\; = 1 \qquad\qquad\quad \text{neutron} \quad N \equiv pnn\end{aligned}$$

hence,

$$\mu_N = \tfrac{1}{2}[(g_p + g_n) \pm \tfrac{5}{3}(g_p - g_n)]\langle S_{zN}\rangle \tag{7.2.30}$$

$$+ \text{ for } N \equiv ppn \qquad - \text{ for } N \equiv pnn.$$

Finally if the quark $g$ factors are assumed to be proportional to their charges (7.2.12) then $g_p = -2g_n$, thus, upon reverting to subscripts $p$ and $n$ for the physical proton and neutron, we find

$$\mu_p/\mu_n = -1{\cdot}5.$$

Experiment gives $-1{\cdot}46$; on the other hand, attempts to apply a similar argument to $\beta$-decay [§ 6.2(b)] gives

$$G_V/G_A = \tfrac{5}{3}$$

as the ratio of the vector and axial vector coupling strengths whereas experiment gives $1{\cdot}18$.

Many interesting mass relations between particles can also be established, by simply assuming that the $\Lambda$ quark is heavier than the proton and neutron quarks and then adding up constituents. For example, the relation

$$M_a = M_m(1 + \sigma Y)$$

for the $\frac{3}{2}^+$ decuplet [§ 7.2(d)] automatically follows from the above assumption. However, the assumption also leads to the requirement $M_\Sigma = M_\Lambda$ in the $\frac{1}{2}^+$ octet, and there does not appear to be any simple way of avoiding this discrepancy.

To finally sum up, some higher symmetry system exists as shown by the multiplets 1, 8 and 10, but it is badly broken as evidenced by the mass spectrum within the multiplets. $SU_3$ provides a reasonable description, and can often be considerably simplified and in places some wrong predictions avoided with the aid of quarks. On the other hand, quark schemes on their own can occasionally lead to wrong predictions. It remains to be seen whether quarks can be found by experiment, or whether they are temporarily useful just as the ether was in the nineteenth century.

### 7.3. Interactions at High Energies

Previously we examined the partial wave technique and obtained the scattering amplitude (7.1.6)

$$f(\theta) = \sum_l (2l + 1)\frac{\eta_l - 1}{2ik} P_l^0(\theta)$$

and the cross-sections

$$\frac{d\sigma}{d\Omega} = |f(\theta)|^2 \qquad \sigma_{sc} = \frac{\pi}{k^2}\sum_l (2l + 1)\,|\eta_l - 1|^2$$

$$\sigma_r = \frac{\pi}{k^2}\sum_l (2l + 1)\,(1 - |\eta_l|^2).$$

When the laboratory momentum of the incident particles is greater than ~5 GeV, the cross-sections are observed to be much smoother as a function of energy than at lower energies (Fig. 7.16), and an alternative parameterisation to that of $s$-channel resonances is useful in the above equations.

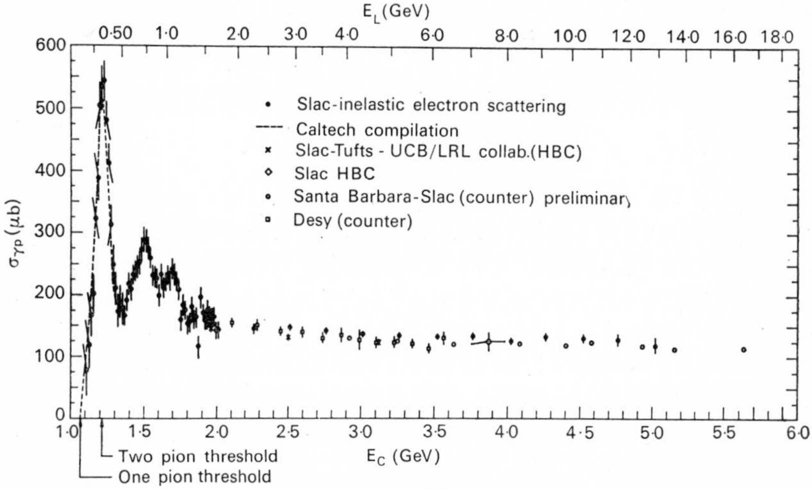


FIG. 7.16. Total cross-section for the interaction of photons with protons as a function of energy (Diebold, 1969).

A convenient first approach to the high-energy data is to consider the problem semiclassically. We can then represent the angular momentum as

$$\mathbf{L} = \mathbf{r} \times \mathbf{k}.$$

Thus if we assume that we have a radius of interaction $R$ (a good guess for strong interactions would be $m_\pi^{-1}$) then particle waves up to a maximum

$$l_{\max} = kR$$

would be expected to take part in the interaction.

If we assume that incoming partial waves up to $l_{\max}$ are completely absorbed we can then write

$$\eta_l = 0 \qquad l < kR$$

$$\eta_l = 1 \qquad l > kR.$$

This is the "black body" approximation and leads to

$$\sigma_{sc} = \sigma_r = \frac{\pi}{k^2} \sum_{l=0}^{kR} (2l + 1) \tag{7.3.1}$$

now

$$\sum_{l=0}^{kR} (2l + 1) \sim k^2 R^2$$

and so

$$\begin{aligned} \sigma_{sc} &= \sigma_r = \pi R^2 \\ \sigma_T &= \sigma_{sc} + \sigma_r = 2\pi R^2. \end{aligned} \tag{7.3.2}$$

Data on the measured total cross-sections at high energies (Allaby *et al.*, 1969) are displayed in Fig. 7.17. It can be seen that the meson–nucleon cross-sections do appear to be approaching constant values at high energies. On the other hand, the above relation would lead to the expectation that the ratio of the elastic scattering to total cross-section is $\frac{1}{2}$; experimentally the ratio is much lower—of the order $\frac{1}{4}$ at ~20 GeV and still apparently dropping.

Consider next the scattering amplitude at high energies. Our approximation implies that

$$f(\theta) = \frac{i}{2k} \sum_{l=0}^{kR} (2l + 1) P_l^0(\theta) \tag{7.3.3}$$

that is a purely imaginary amplitude. This has been checked near $\theta = 0$ by examining the interference with Coulomb scattering; one then has a combination of two amplitudes—the Coulomb amplitude $f_C$ and nuclear $f_N$

$$f_{exp}(\theta) = f_C + f_N = (f_C + \text{Re}\, f_N) + i\, \text{Im}\, f_N$$

$$\frac{d\sigma}{d\Omega} = (f_C + \text{Re}\, f_N)^2 + (\text{Im}\, f_N)^2.$$

The Coulomb amplitude $f_C$ can be calculated in the manner indicated in § 5.1 and Im $f_N$ at $\theta \to 0$ can be determined from the optical theorem (7.1.9), hence Re $f_N$ may be determined. Typical results are shown in Fig. 7.17 (Foley *et al.*, 1967); the data indicates that Re $f_N$ may well approach zero at very high energies.

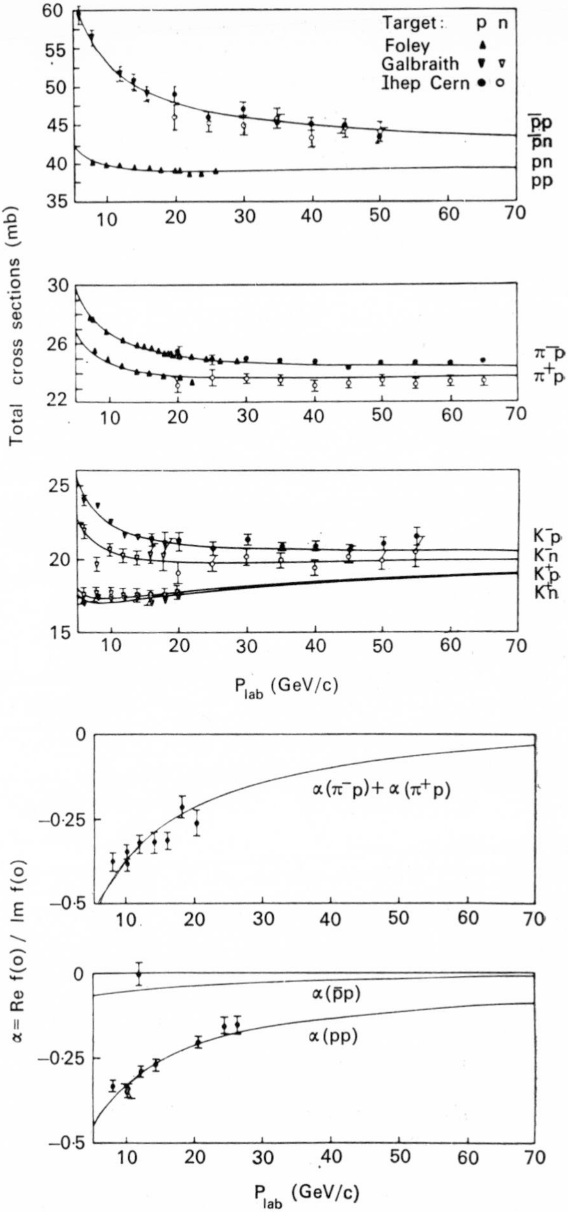


FIG. 7.17. Behaviour of total cross-sections at high energies (Allaby *et al.*, 1969) and the ratio of the real to imaginary forward scattering amplitudes (Foley *et al.*, 1967). The fitted curves are those of Barger and Phillips (1970).

The scattering cross-section

$$\frac{d\sigma}{d\Omega} = |f(\theta)|^2$$

from the black body approximation leads to curves of the type shown in Fig. 7.18, that is a central diffraction peak and pronounced minima and secondary maxima. Experimentally the peak occurs but not the minima; this is not surprising as the large angle behaviour arises from the sharpness of the boundary $R$ chosen for our parameterisation.

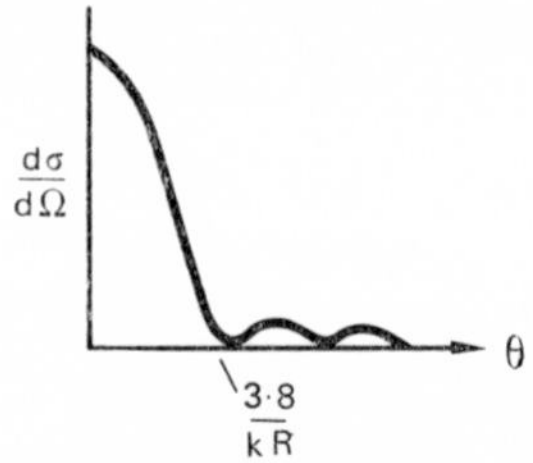


FIG. 7.18. Schematic behaviour of differential scattering cross-section predicted by the black body approximation.

Secondary maxima do appear sometimes in high energy process but the associated dips seem to be associated more with the parameter $t$ (the square of the four momentum transferred) than $\theta$ (Fig. 7.19).

Results of this type are illustrated in Fig. 7.19 for $\pi^- p \to \pi^0 n$; $\log d\sigma/dt$ is plotted versus $t$ for different laboratory momenta. If we neglect the difference in mass between $\pi^-$ and $\pi^0$, then by (4.3.1)

$$\begin{aligned} t &= -(k_1 - k_3)^2 \\ &= -2k^2(1 - \cos\theta) \quad \text{in } c\text{-system} \end{aligned}$$

where $k$ is the linear momentum in the $c$-system, thus

$$\frac{d\sigma}{dt} = \frac{-\pi}{k^2}\frac{d\sigma}{d\Omega} \xrightarrow[s\to\infty]{} \frac{-4\pi}{s}\frac{d\sigma}{d\Omega}. \tag{7.3.4}$$

The black body approximation is capable of many refinements, and they generally appear under the title of the eikonal model in the literature.

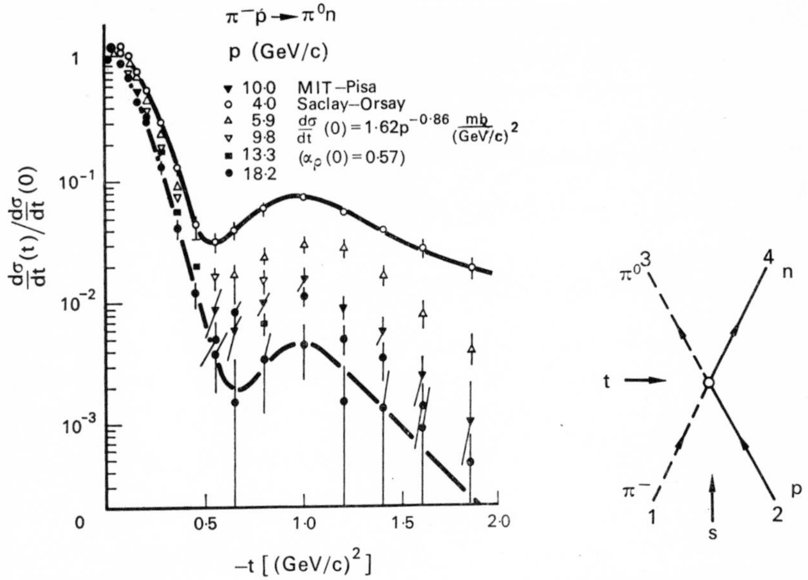


FIG. 7.19. Cross-section as a function of $t$ for $\pi^- p \rightarrow \pi^0 n$ (report by van Hove, 1966).

However, the method is essentially confirmed to elastic and total cross-sections and makes no provision for reactions of the type $ab \rightarrow cd$ where $c$ and $d$ are different from $a$ and $b$. Some more general approach at high energies is desirable and this will be discussed in subsequent sections.

## 7.4. Regge Poles

Earlier in this chapter we have seen how peaks in the scattering cross-sections at low energies can be related to resonating amplitudes in the partial wave expansion, and in turn to the formation of resonant particle states in the $s$-channel. At high energies the peaks disappear and at the same time the angular distribution becomes strongly peaked in the forward direction (Figs. 7.18 and 7.19). This behaviour is characteristic of particle exchanges in the $t$-channel (compare $e^- e^-$ scattering and Fig. 5.2 in § 5.1). Thus it is tempting to relate the black body approximation of § 7.3 to particle exchange models using the Feynman rules. However, a grave difficulty arises with this approach as we shall see later in this section. Before this problem and a possible solution (Regge poles) are examined, the concept of crossing symmetry must be introduced.

### 7.4(a). Crossing symmetry

Consider a two-body process involving spinless particles

$$\underset{1}{\pi^+}\,\underset{2}{\pi^+} \rightarrow \underset{3}{\pi^+}\,\underset{4}{\pi^+}.$$

We have two final four momenta, i.e. eight parameters to describe the interaction, the Einstein relation $E^2 = \mathbf{p}^2 + m^2$ reduces these to six and $\delta(p_i - p_f)$ to two.‡ We shall choose these as the Mandelstam variables $s$, $t$ (4.3.1)

$$T = T(s, t)$$

where

$$s = -(k_1 + k_2)^2 = E_c^2$$

$$t = -(k_1 - k_3)^2 = -2k^2(1 - \cos\theta_s) \quad \text{in } c\text{-system}$$

$$u = -(k_1 - k_4)^2$$

$$s + t + u = 4m^2 \qquad m = m_\pi.$$

Now consider the situation when we charge conjugate particles 2 and 4

$$\underset{1}{\pi^+}\,\underset{\bar{2}}{\pi^-} \rightarrow \underset{3}{\pi^+}\,\underset{\bar{4}}{\pi^-}$$

and simultaneously reverse their four momenta

$$k_2 = -k_{\bar{2}} \qquad k_3 = -k_{\bar{3}}. \tag{7.4.1}$$

We then have a new amplitude $\bar{T}(\bar{s}, \bar{t})$. We can then represent the situation by the diagrams in Fig. 7.20.

The kinematic variables associated with this figure are

$$s = -(k_1 + k_2)^2 \qquad \bar{s} = -(k_1 + k_{\bar{3}})^2 = -(k_1 - k_3)^2 = t$$

$$t = -(k_1 - k_3)^2 \qquad \bar{t} = -(k_1 - k_{\bar{2}})^2 = -(k_1 + k_2)^2 = s.$$

‡ Recall that $\delta(p_i - p_f)$ is a product of four delta functions (2.2.12).

The crossing symmetry principle then asserts that the amplitude remains the same in form in going from particle to antiparticle providing we change the sign of the four momentum—this is also known as the substitution rule. Thus we find

$$T(s, t) = \bar{T}(\bar{s}, \bar{t}) = \bar{T}(t, s). \tag{7.4.2}$$

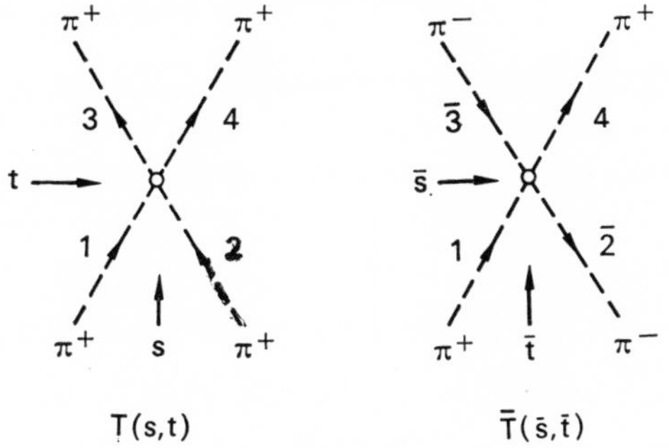


FIG. 7.20.

Thus $t$ and $s$ have changed roles in the two amplitudes ($T$ and $\bar{T}$ are often called the $s$- and $t$-channel amplitudes respectively). The correctness of the assertion can be shown in specific examples; for example, if we

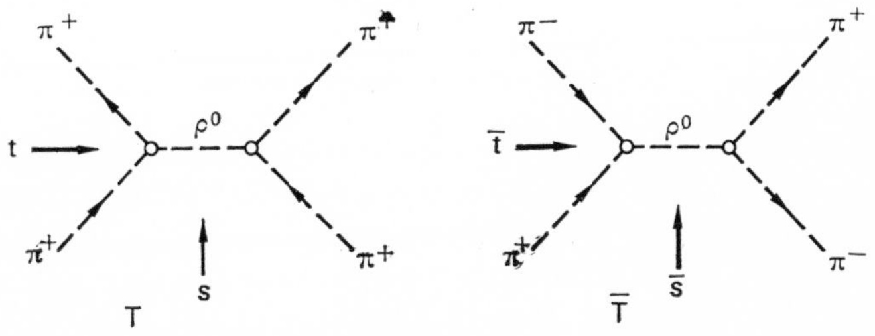


FIG. 7.21.

assume $\varrho$ dominance (Fig. 7.21) and evaluate the $\pi\pi$ interaction in this approximation, then we have the situation illustrated in Fig. 7.21. It is a straightforward task to write down the amplitudes for the two processes and show that $T = \bar{T}$ when evaluated with the kinematic conditions (7.4.1).‡

Notice that the physical regions are not the same for $T$ and $\bar{T}$. The limits for the physical regions for $s$ and $t$ are

| $T$ | $\bar{T}$ |
|---|---|
| $s \geqq 4m^2 \quad t \leqq 0$ | $s \leqq 0 \quad t \geqq 4m^2$ |
| *s*-channel | *t*-channel |

Thus the two regions do not overlap (Fig. 7.22) and we are asserting that $T(s, t)$ is an analytic function of its arguments in order that continuation from one physical region to another is possible.

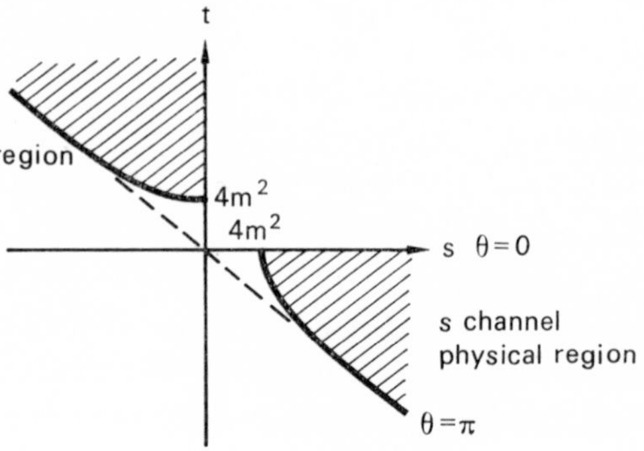


FIG. 7.22.

### 7.4(b). Problems with one-particle exchange models

When we examined the partial wave expansion we found that the form of the propagator for an $s$-channel resonance involving a system with spin $j$ could be represented by equation (7.1.17)

$$\frac{g^2 P_j(\cos\theta_s)}{(m_j^2 - s) - im_j\Gamma} \tag{7.4.3}$$

where $\cos\theta_s$ is given by

$$t = -(k_1 - k_3)^2 = -2k^2(1 - \cos\theta_s) \quad \text{in } c\text{-system}$$

‡ Another occurrence of the crossing principle can be seen in the amplitudes for $e\pi \to e\pi$ scattering (5.3.9) and $e^+e^- \to \pi^+\pi^-$ (5.3.12).

hence

$$\cos\theta_s = 1 + \frac{t}{2k^2}$$

and if we assume the initial particles $a$ and $b$ are of equal mass $m$

$$s = E_c^2 = 4(k^2 + m^2)$$

$$\cos\theta_s = 1 + \frac{t}{2k^2} = 1 + \frac{2t}{s - 4m^2} \tag{7.4.4}$$

Similarly for a $t$-channel amplitude with particles of mass $m$

$$\cos\theta_t = 1 + \frac{2s}{t - 4m^2}. \tag{7.4.5}$$

Now consider a scattering process involving the exchange of a particle of spin $j$ in the $t$-channel (Fig. 7.23).

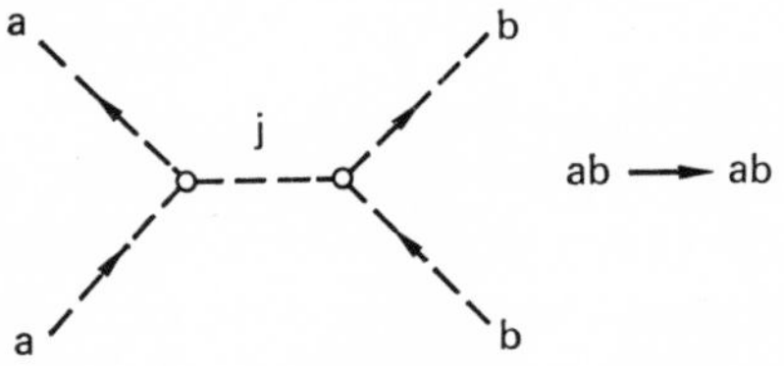


FIG. 7.23.

In the physical region for the $t$-channel amplitude ($t > 4m^2$) we can use our crossing symmetry argument [§ 7.4(a)] to represent the amplitude as

$$T_{fi} \propto \frac{g^2 P_j(\cos\theta_t)}{(m_j^2 - t) - im_j\Gamma}.$$

Now $P_j$ (cos $\theta_t$) is a polynomial in cos $\theta_t$

$$P_j(\cos\theta_t) = \sum_{i=0}^{i=j} a_i(\cos\theta_t)^i$$

and so in the physical region for the $s$-channel with conditions

$$s \to \infty \qquad -t \to 0$$

then from (7.4.5)

$$\cos\theta_t \to \frac{-s}{2m^2}$$

and only the highest-order term is of importance in the polynomial

$$P_j(\cos\theta_t) \to (\cos\theta_t)^j \to \left(\frac{-s}{2m^2}\right)^j. \qquad (7.4.6)$$

Hence

$$|T_{fi}|^2 \propto s^{2j}$$

and from (7.3.4) and (2.4.3)

$$\frac{d\sigma}{dt} = \frac{-4\pi}{s}\frac{d\sigma}{d\Omega} \propto \frac{-4\pi}{s}\frac{1}{E_c^2}|T_{fi}|^2 \propto s^{2j-2}. \qquad (7.4.7)$$

Thus if $j > 2$ the one-particle exchange model predicts $d\sigma/dt \to \infty$ as $s \to \infty$. This conclusion has alarming theoretical implications and is contrary to experiment.

### 7.4(c). Regge poles

These arose from the work of Regge (1959, 1960), on non-perturbative solutions of the Schrodinger equation for a wide range of potentials. He argued that the scattering amplitude could be an analytic (i.e. continuous) function of angular momentum $j$ as well as of energy. The idea was taken up by Chew and his collaborators (1961, 1962) who conjectured that it could be applied to high-energy processes, although rigorous proofs of this have never been given.

The Regge pole model at high energies retains some of the features of the Feynman graph technique (§ 4.4). As in that method the assumption is made that the scattering amplitude is dominated by particle exchange in the $t$-channel. The exchanged systems have the appropriate internal quantum numbers ($G$ parity, strangeness, etc.) to satisfy the conservation laws at the vertex, but in contrast to the Feynman rules the requirement of the exchange of particles of fixed spin is relaxed. Instead of assuming, say, the exchange of a pion (spin 0), one assumes the exchange of a system which has effective spin zero only at the point $t = m_\pi^2$—a point in the unphysical region for elastic scattering ($t \leqq 0$).

Consider the elastic scattering of spinless particles $ab \to ab$ in the $s$-channel. If we first work in the $t$-channel we can represent the amplitude as

$$T(t, \cos\theta_t) = \sum_j A(t, j)\, P_j(\cos\theta_t) \tag{7.4.8}$$

where $A(t, j)$ contains constants, phase shifts and kinematic factors. We next cease to regard $j$ as a series of integers as in the partial wave expansion by replacing it by a continuous variable

$$j \to \alpha(t) \qquad \alpha \text{ complex} \tag{7.4.9}$$

so that $T$ can be analytically continued in the complex $j$-plane. Now using a mathematical trick, known as the Watson–Sommerfeld transformation, we can rewrite equation (7.4.8) as a contour integral

$$T(t, \cos\theta_t) = \frac{1}{2\pi i} \oint d\alpha\, A(t, \alpha) \frac{P_\alpha(-\cos\theta_t)}{\sin\pi\alpha}$$

where the contour just surrounds the Re $\alpha$-axis (Fig. 7.24). It can easily be seen that the above expression reduces to (7.4.8) by noting that

$$\underset{\alpha \to j.}{\text{residue}} \frac{\pi}{\sin\pi\alpha} = (-1)^j \text{ and } (-1)^j P_j(-\cos\theta_t) = P_j(\cos\theta_t)$$

The proof of the first equation follows from writing $\alpha = j + \delta$ and letting $\delta \to 0$; the second follows from the properties of the Legendre polynomials.

Regge showed that for a wide range of potentials the contour could be distorted to the line $\alpha = -\frac{1}{2} - i\infty$ to $\alpha = -\frac{1}{2} + i\infty$ without encountering any singularities other than the poles in $A(t, \alpha)$ ($\alpha_1$ and $\alpha_2$ in Fig. 7.24). Thus, using the rules of contour integration we can write the amplitude as

$$T(t, \cos\theta_t) = \frac{-1}{2\pi i} \int_{-\frac{1}{2}-i\infty}^{-\frac{1}{2}+i\infty} d\alpha \frac{A(t, \alpha)\, P_\alpha(-\cos\theta_t)}{\sin\pi\alpha} + \sum_i \phi_i(t) \frac{P_{\alpha i}(-\cos\theta_t)}{\sin\pi\alpha_i} \tag{7.4.10}$$

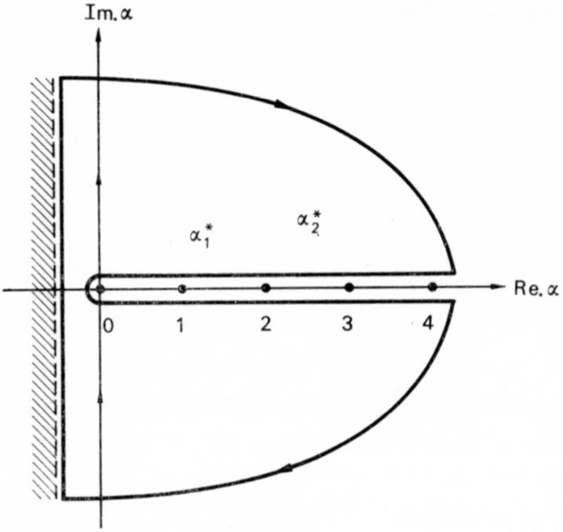


FIG. 7.24.

where $\alpha_i$ is the $i$th pole in $\alpha$ and $\phi_i$ is the residue at $\alpha_i$ (apart from the polynomial).

Consider the pole term; if $\alpha = \alpha(t)$ passes through a resonance $\text{Re}\,\alpha = j$ at $t = m_j^2$, then in the region of the pole (where $j$ is an integer) for small $\text{Im}\,\alpha$

$$\alpha(t) = j + \alpha'(t - m_j^2) + i\,\text{Im}\,\alpha$$

$$\sin \pi\alpha = (-1)^J \pi[\alpha'(t - m_j^2) + i\,\text{Im}\,\alpha]$$

where $\alpha' = d\alpha/dt$. Thus the pole term becomes

$$\frac{\phi(t)\, P_j(\cos\theta_t)}{\pi\alpha'\left(t - m_j^2 + i\dfrac{\text{Im}\,\alpha}{\alpha'}\right)}$$

and we have the same basic format as in § 7.4(b).

Next let us examine how the amplitude $T(t, \cos\theta_t)$ behaves in the physical region of the $s$-channel at high energies. If we have

$$s \to \infty \qquad -t \to 0$$

then from (7.4.5) and (7.4.6)

$$\cos\theta_t \to \frac{-s}{2m^2} \qquad P_\alpha(-\cos\theta_t) \to P_\alpha\left(\frac{s}{2m^2}\right) \propto s^\alpha$$
$$s \to \infty \qquad\qquad s \to \infty.$$

In these circumstances it can be shown that the integral in (7.4.10) tends to zero at least as fast as $(\cos\theta_t)^{-\frac{1}{2}}$, i.e. as $s^{-\frac{1}{2}}$, hence $T$ is dominated by the pole terms and we have

$$T(\cos\vartheta_t, t) = T(s, t) = \sum_i \frac{\gamma_i(t)}{\sin\pi\alpha_i}\left(\frac{s}{s_0}\right)^{\alpha_i} \tag{7.4.11}$$

where we have introduced a term $s_0$ for convenience of scaling (generally one takes $s_0 = 1\ \text{GeV}^2$) and redefined our residues accordingly.

Now $\alpha$ is a continuous function of $t$

$$\alpha = \alpha(t)$$

and its path is referred to as a Regge trajectory. The important difference between the Regge approach and our previous method using arguments based on the exchange of a particle of fixed spin now emerges. In the latter treatment we had a fixed pole at some point $t = m^2$; now we have a series of continuously moving poles. The cross-sections $d\sigma/dt$ at high energies are [compare (7.4.7)]

$$\frac{d\sigma}{dt} \propto s^{2j-2} \qquad \text{one particle exchange}$$
$$s^{2\alpha-2} \qquad \text{Regge} \tag{7.4.12}$$

and it is evident that provided

$$\alpha < 1 \quad \text{for} \quad t < 0 \qquad \text{in } s\text{-channel}$$

the divergence difficulties are avoided.

### 7.4(d). Behaviour of the Regge amplitude

Assume for the moment that we have only one Regge pole term contributing to the Regge amplitude, then from equation (7.4.11)

$$\frac{d\sigma}{dt} = F(t)\left(\frac{s}{s_0}\right)^{2\alpha-2}$$

and so

$$\log \frac{d\sigma}{dt} = \log F(t) + (2\alpha - 2) \log\left(\frac{s}{s_0}\right). \quad (7.4.13)$$

Thus a plot of log $d\sigma/dt$ for fixed $t$ as a function of log $s$ gives the value of $\alpha$ for that value of $t$. The plot is repeated for different $t$ values and hence $\alpha(t)$ determined.

The problem is to find a relatively clean amplitude to compare with experiment, since many pole terms can often contribute to the same amplitude. One of the cleanest appears to be $\pi^- p \to \pi^0 n$ (Fig. 7.25).

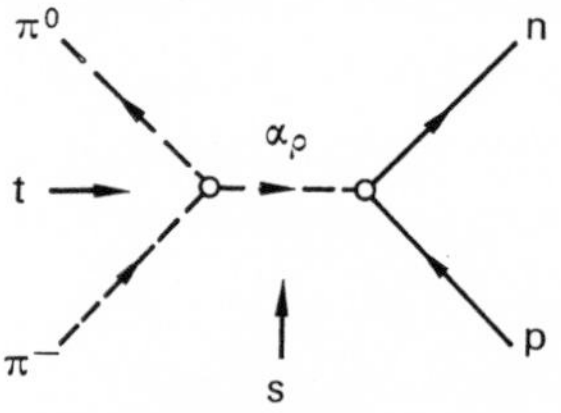


FIG. 7.25.

In terms of Feynman diagrams the simplest system which could be exchanged would be a $\varrho$-meson since this has all the right quantum numbers, $j^P = 1^-$, $I = 1$, $G = (-1)^n = +1$. The Regge trajectory is therefore denoted as $\alpha_\varrho$. The basic experimental data to be analysed is shown in Fig. 7.19 and when analysed using (7.4.13) gives a simple straight-line relation (the units are $\text{GeV}^{-2}$ for $\alpha'$)

$$\alpha(t) = \alpha(0) + \alpha' t = 0{\cdot}57 + 0{\cdot}91\, t. \quad (7.4.14)$$

For an elastic scattering process in the $s$-channel the physical region of $t$ is $\leqq 0$. If we extrapolate $\alpha_\varrho$ to $t > 0$, i.e. the $t$-channel region (Fig. 7.26), then $\alpha_\varrho$ passes through $t = m_\varrho^2 = 0{\cdot}76^2\ \text{GeV}^2$ at $\alpha_\varrho = 1{\cdot}09$ which is close to the spin value of 1 for the $\varrho$-meson. Furthermore, at $t = 1{\cdot}6^2\ \text{GeV}^2$ it goes through $\alpha_\varrho = 3$. Now there is a known meson with mass 1·6 GeV, the $g$-meson, which decays $g \to 2\pi$,‡ and although

‡ The $2\pi$ decay mode of the $g$-meson limits it to spin-parity $0^+, 1^-, 2^+, 3^-, \ldots$ from (6.1.5).

the value is still not completely certain the present experimental data indicated a spin of 3.

Now consider why $\alpha_\varrho \equiv j = 1, 3$ was mentioned but not $j = 2$. The reason arises from the requirement of Bose–Einstein symmetry. According to Fig. 7.25 we must exchange a system with charge $-e$ and $I_3 = -1$;

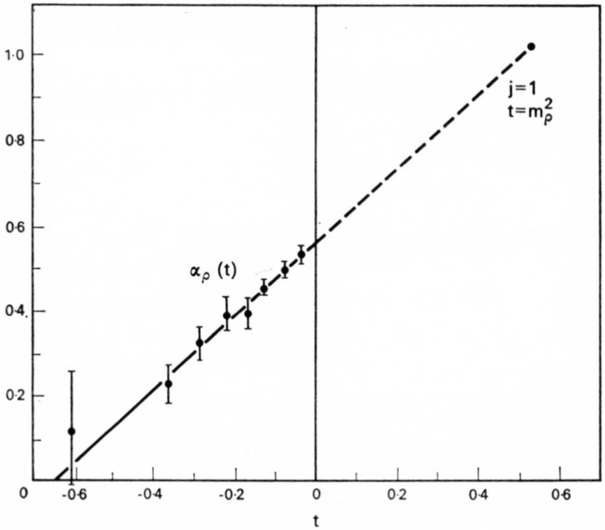


FIG. 7.26. Rho trajectory deduced from the reaction $\pi^- p \to \pi^0 n$ (report by van Hove, 1966).

since $I = 2$ mesons do not appear to exist we are therefore limited to an $I = 1$ system. Pions have isospin $i = 1$, thus the only way we can form an $I = 1$ object from two pions is

$$\mathbf{I} = \mathbf{i}_1 \times \mathbf{i}_2 \equiv \mathbf{1}.$$

This is antisymmetric under exchange of 1 and 2 and so must be combined with an antisymmetric space function [see the analysis of spin-parity in § 7.1(e)]. For two pions spatial exchange is formally similar to the parity operation (6.1.5) and so $P_{ex} = (-1)^l = (-1)^j$; thus we have $j = 1, 3, 5 \ldots$.

In Regge theory this behaviour is allowed for by introducing a signature factor; the form for this factor is

$$\zeta(t) = \tfrac{1}{2}\,(1 + \tau\, e^{-i\pi\alpha(t)}) \tag{7.4.15}$$

and for $\tau = -1$

$$\begin{aligned} \zeta(t) &= 0 \qquad \alpha(t) \equiv j = 0, 2, 4 \ldots \\ &= 1 \qquad \alpha(t) \equiv j = 1, 3, 5 \ldots \end{aligned}$$

with the reverse for $\tau = +1$.

In general for two external particles of parity $P_1$ and $P_2$ and a trajectory with poles of parity $P_i$

$$\tau_i = P_1 P_2 P_i .$$

One further modification is desirable for the Regge amplitude. The integration in Fig. 7.24 was continued as far as $\mathrm{Re}\,\alpha = -\frac{1}{2}$. It is desirable to carry it further into the negative region, but unwanted poles would then appear for $\alpha = -1, -2, -3 \ldots$; they can be excluded by including a factor $1/\Gamma(\alpha)$ since the gamma function $\Gamma \to \infty$ at $\alpha = 0, -1, -2 \ldots.$ The Regge amplitude (7.4.11) thus becomes

$$\begin{aligned} T(s, t) &= \sum_i \frac{\gamma_i(t)}{\sin \pi\alpha_i} \left(\frac{s}{s_0}\right)^{\alpha_i} \\ &\to \sum_i \frac{\beta_i(t)}{\Gamma(\alpha_i)} \, \frac{1 + \tau_i\, e^{-i\pi\alpha_i}}{\sin \pi\alpha_i} \left(\frac{s}{s_0}\right)^{\alpha_i} \end{aligned} \tag{7.4.16}$$

where we have absorbed the $\frac{1}{2}$ in (7.4.15) into the $\beta_i(t)$.

Regge amplitudes can also be constructed for fermion exchange. The signature factors must again satisfy invariance requirements and lead to pole terms at intervals of $\Delta j = 2$.

### 7.4(e). Successes and failures

In Fig. 7.19 a dip appears at $t \sim -0{\cdot}6$ GeV$^2$ in the process $\pi^- p \to \pi^0 n$. A simple explanation is offered by Regge pole theory. Consider the crossed channel $\pi^-\pi^0 \to \bar{p}n$ which we assume is dominated by the $\varrho$-trajectory.

The amplitude in this channel can be represented as a sum of two terms, depending on whether the $\bar{p}n$ system is in a singlet or triplet spin state

$$T_{fi} \sim CP_\alpha^0(\cos\theta_t) + DP_\alpha^1(\cos\theta_t). \tag{7.4.17}$$

This amplitude is no more than a statement of conservation of angular momentum, $P^1$ must combine with the triplet spin function to give the same angular momentum conditions as in the initial state. Now the associated Legendre polynomial can also be written as

$$P_\alpha^1(\cos\theta_t) = -\sin\theta_t \frac{\partial}{\partial(\cos\theta_t)} P_\alpha^0(\cos\theta_t)$$

If we now consider the physical region of the $s$-channel $\pi^- p \to \pi^0 n$ for $s \to \infty$ [compare (7.4.5)]

$$\cos\theta_t \propto s \quad \sin\theta_t = \sqrt{1-\cos^2\theta_t} \sim \cos\theta_t \propto s$$

$$P_\alpha^1(\cos\theta_t) \propto s\frac{\partial}{\partial s} s^\alpha = \alpha s^\alpha$$

and equation (7.4.17) becomes

$$T_{fi} \sim [A(t) + \alpha B'(t)]\left(\frac{s}{s_0}\right)^\alpha$$

$$= [A(t) + \alpha\sqrt{-t}\, B(t)]\left(\frac{s}{s_0}\right)^\alpha$$

where the $\sqrt{-t}$ was extracted from $B'(t)$ since spin flip terms vanish as $\theta_s \to 0$, i.e. as $\sqrt{-t}$ tends to zero [compare (7.1.20) and the definition of $t$ in § 7.4(a)].

An inspection of $T_{fi}$ shows that the second term should vanish at $\alpha = 0$, and if we recall the form of the trajectory (7.4.14)

$$\alpha_\varrho = 0{\cdot}57 + 0{\cdot}91\, t$$

then $\alpha_\varrho = 0$ at $t \sim -0{\cdot}6$ GeV$^2$, and so a dip in $\pi^- p \to \pi^0 n$ might be expected at this point and is seen in Fig. 7.19.

The behaviour of the $\varrho$ trajectory (Fig. 7.26) and the explanation of the dip at $t = -0{\cdot}6$ GeV$^2$ in $\pi^- p \to \pi^0 n$ scattering represent probably the most successful applications of the Regge theory. Against them must be set a considerable number of failures. As a first example the dip at $t = -0{\cdot}6\ \text{GeV}^2$ should be accompanied by zero polarisation of the recoiling neutron if the theory is considered in a little more detail. Experimentally the polarisation is ~15 per cent.

Dips are also found in backward $\pi p$ scattering as well as in forward scattering. This is illustrated for $\pi^+p$ scattering in Fig. 7.27. Backward

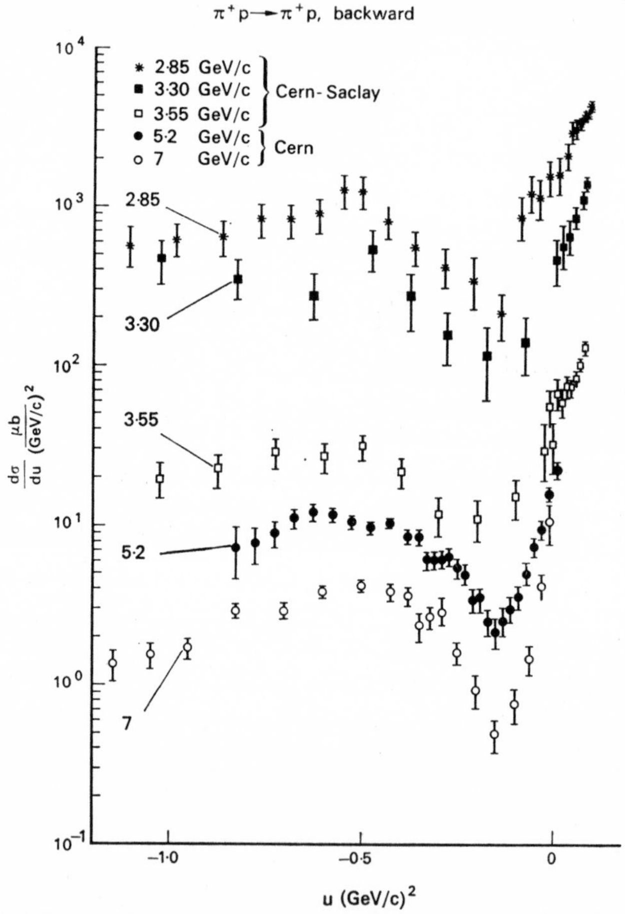


FIG. 7.27. Backward $\pi^+p$ scattering as a function of $u$ (report by Belletini, 1968).

scattering is dominated by baryon exchange in the $u$-channel, instead of meson exchange in the $t$-channel as illustrated in Fig. 7.28 (compare also the effect of $t$ and $u$ channel exchanges discussed for electron–electron scattering in § 5.1). Backward $\pi^+p$ scattering can arise both from the

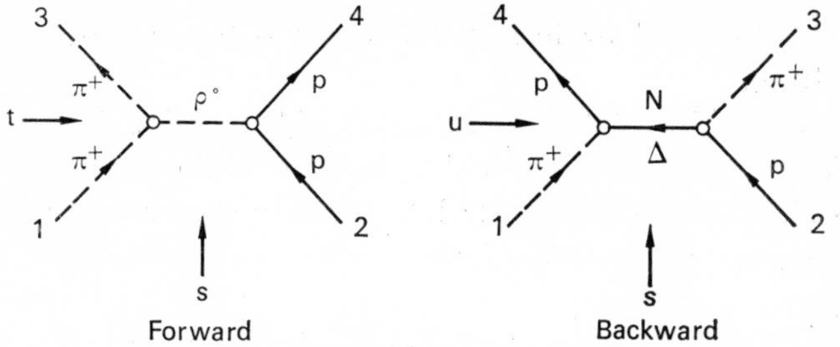


FIG. 7.28.

exchange of isospin $\frac{1}{2}(N)$ and $\frac{3}{2}(\Delta)$ nucleons. An impressive correlation was obtained by Barger and Cline (1968), between the Regge trajectories obtained in fitting the scattering data ($u < 0$) and the observed behaviour of nucleon resonance masses as a function of spin ($u > 0$); the result is shown below:

| | scattering data | observed resonances | |
|---|---|---|---|
| | $u < 0$ | $u > 0$ | |
| $\alpha_N =$ | $-0{\cdot}38 + 0{\cdot}88\,u$ | $-0{\cdot}34 + 1{\cdot}0\,u$ | for $N$ |
| $\alpha_\Delta =$ | $+0{\cdot}19 + 0{\cdot}87\,u$ | $+0{\cdot}15 + 0{\cdot}9\,u$ | for $\Delta$. |

However, the Regge analysis also predicted a dip at $u = -1{\cdot}9$ GeV$^2$ in $\pi^-p$ elastic scattering and this was not found.

A further unsatisfactory situation from the point of view of Regge theory appears to exist in photoproduction. Here fixed poles, i.e. poles which do not move as a function of $t$ or $u$, appear to exist. If the experimental data is parameterised as

$$\frac{d\sigma}{dt} = F(t)\, s^{2\alpha-2}$$

where we forget about the subtleties of more than one trajectory exchanged and regard $\alpha$ as a single parameter, then the data for the processes

$$\gamma p \to \pi^0 p \qquad \text{forward direction}$$

$$\gamma p \to \pi^+ n \qquad \text{backward direction}$$

can be represented by fixed poles with $\alpha \sim 0$ and $-0{\cdot}5$ respectively (Fox, 1969).

Lastly consider total cross-sections at high energies. As we have seen in § 7.3 these appear in general to be approaching constant values as $s \to \infty$. The optical theorem (7.1.9) relates total cross-section $\sigma_T$ to the imaginary part of the forward elastic scattering amplitude (i.e. the point $t = 0$)

$$\sigma_t \propto \frac{1}{kE_c} \operatorname{Im} T_{ii} \xrightarrow[s \to \infty]{} \frac{1}{s} \operatorname{Im} T(s, t = 0).$$

Thus if we consider the basic form of the Regge amplitude (7.4.11) we find

$$\sigma_T \propto \frac{1}{s} \operatorname{Im} T(s, t = 0) \propto s^{\alpha(0)-1}.$$

It is apparent that if the cross-section is dominated by a single trajectory, then a constant cross-section implies that $\alpha(0) = 1$. No known trajectory passes through this point; nevertheless, the existence of such a trajectory is conjectured. It is known as the pomeron trajectory and must have zero internal quantum numbers since no states are altered in forward elastic scattering. The behaviour of the diffraction peak in forward elastic scattering implies that the slope of the pomeron trajectory is small ($\alpha' < 0{\cdot}5$ GeV$^2$); this behaviour is at variance with all other known trajectories which have $\alpha' \sim 1{\cdot}0$ GeV$^2$. The pomeron may well play the same role in high-energy scattering as quarks do in higher symmetry schemes—both could be semi-mythological creatures on which to fix ideas.

## 7.5. Finite Energy Sum Rules

So far in this chapter the regions of low- and high-energy scattering have been treated as separate and unrelated problems. In each region a different parameterisation was used, and no connection was apparent between the $s$-channel resonance (low energy) and the $t$-channel Regge (high energy) parameters. Obviously there is at least a region of overlap at intermediate energies (see § 7.6). The two regions can in fact be connected more formally by means of the analytic properties of the scattering amplitude, and the associated equations which display this connection are known as the finite energy sum rules (Igi, 1962; Logunov *et al.*, 1967; Dolen *et al.*, 1968).

The basic equation used to express the analytic properties of the amplitude is the following form of the Cauchy equation for complex variables:

$$A(\nu) = \frac{1}{\pi} \int_{-\infty}^{\infty} d\nu' \frac{\operatorname{Im} A(\nu')}{\nu' - \nu} \tag{7.5.1}$$

[compare (4.3.8)], where $A$ is the scattering amplitude after kinematic terms have been removed [a specific example is given in (7.7.1)], and the variable $\nu$ is

$$\nu = \frac{s - u}{4m} = E_L + \frac{t}{4m}.$$

Thus $A$ is an analytic function of the laboratory energy $E_L$ of the incident particle if we assume that the four-momentum transfer $t$ is kept fixed. Equation (7.5.1) is also called a dispersion relation.

An amplitude which is symmetric (antisymmetric) under crossing ($s \leftrightarrow u$) is symmetric (antisymmetric) in $\nu \leftrightarrow -\nu$. An antisymmetric amplitude will be chosen for our present discussion

$$A^*(\nu) = -A(-\nu) \tag{7.5.2}$$

so that

$$\operatorname{Im} A(\nu) = \operatorname{Im} A(-\nu) \tag{7.5.3}$$

but this choice is not essential.

Following Logunov *et al.* (1967) we assume that $A(\nu)$ is of the form

$$A(\nu) = \sum_i C_i \nu_i^{\alpha_i} + \varepsilon(\nu) \tag{7.5.4}$$

where $\varepsilon(\nu)$ decreases as $\nu \to \infty$, i.e. infinite energy, and is negligible for $\nu > N$. Thus at high energies $A(\nu)$ behaves like a sum of Regge poles. If we apply the Cauchy formula (7.5.1) to $\varepsilon(\nu)$ together with crossing symmetry relation (7.5.3), we find

$$\varepsilon(\nu) = \frac{1}{\pi} \int_{-\infty}^{\infty} d\nu' \frac{\operatorname{Im} \varepsilon(\nu')}{\nu' - \nu} = \frac{2\nu}{\pi} \int_{0}^{N} d\nu' \frac{\operatorname{Im} \varepsilon(\nu')}{\nu'^2 - \nu^2}$$

where we have neglected the region

$$\int_{N}^{\infty} d\nu' \operatorname{Im} \varepsilon(\nu').$$

We may now insert this formula in (7.5.4) at the point $\nu = N$ so that $\varepsilon(N) = 0$ and so

$$0 = \frac{2\nu}{\pi} \int_{0}^{N} d\nu' \frac{\operatorname{Im} A(\nu') - \sum_i \operatorname{Im} C_i \nu_i'^{\alpha_i}}{\nu'^2 - \nu^2}$$

Thus the numerator of the integrand is zero, and

$$\int_{0}^{N} d\nu \operatorname{Im} A(\nu) = \sum_i \frac{\operatorname{Im} C_i}{\alpha_i + 1} N^{\alpha_i + 1}. \tag{7.5.5}$$

This is the zeroth order finite energy sum rule. A similar calculation using the expression $\nu^n A(\nu)$ leads to the $n$th order finite energy sum rule

$$\frac{1}{N^{n+1}} \int_{0}^{N} d\nu\, \nu^n \operatorname{Im} A(\nu) = \sum_i \frac{\operatorname{Im} C_i}{\alpha_i + n + 1} N^{\alpha_i + 1}. \tag{7.5.6}$$

The sum rules are also sometimes referred to as superconvergence relations. Equations (7.5.5) and (7.5.6) are consistency conditions relating the amplitude in the low-energy region $0 < \nu < N$ to that for $\nu > N$. The left-hand side can be evaluated in terms of phase shift analysis, or alternatively by summing over a string of resonances since $\operatorname{Im} A(\nu)$ is maximal at a resonance, and the right-hand side can be treated by Regge amplitudes. Assuming their correctness one side of the equations may also be used to determine the parameters of the other side; an example of this technique will be mentioned in § 7.6.

## 7.6. Interference and/or Duality

In § 7.5 mention was made that an area of overlap must exist between the regions of low and high energy scattering and their apparently differing interpretations. Such an area was covered by the work of Kormanyos *et al.* (1966) who made a series of measurements of $\pi^- p$ elastic scattering at 180° as a function of increasing $s$. Their results (Fig. 7.29) showed striking dips, and were interpreted by assuming interference between the two amplitudes associated with the diagrams at the bottom of Fig. 7.29

$$T_{fi} = \underbrace{T\,(\text{Regge in } \alpha_\Delta)}_{\substack{\text{Regge trajectory} \\ \text{in } t\text{-channel}}} + \underbrace{T\,(N^* \text{ in } s\text{-channel})}_{\substack{\text{Resonating partial} \\ \text{wave in } s\text{-channel}}} \tag{7.6.1}$$

(Barger and Cline, 1966). The fit to the experimental results was excellent and was used to assign spin and parity values $7/2^-$ and $11/2^+$ for resonances with masses of 2190 and 2420 MeV respectively.

Despite the satisfactory nature of the fit an uneasy feeling existed that the method of interpretation was not quite right. The statement in (7.6.1)

$$T_{fi} = T_{\text{Regge}} + T_{\text{Resonance}} \tag{7.6.2}$$

("the interference model") does not satisfy the finite energy sum rules of § 7.5, which offer a connection between the direct ($s$) and exchange ($t$ or $u$) channel descriptions of the amplitude (Dolen, Horn and Schmid, 1968).

An alternative viewpoint to that of the interference model is expressed in the concept of duality. This is based upon the principle that both $T_{\text{Regge}}$ and $T_{\text{Resonance}}$ offer complete descriptions of the scattering amplitude, and either one can be used depending on which is simpler and more appropriate in specific situations

$$T_{fi} \sim \langle T_{\text{Regge}} \rangle \sim \langle T_{\text{Resonance}} \rangle \tag{7.6.3}$$

where the symbol $\langle\ \rangle$ implies an average over Regge poles or resonances in a finite region. This statement implies that Regge-like behaviour at

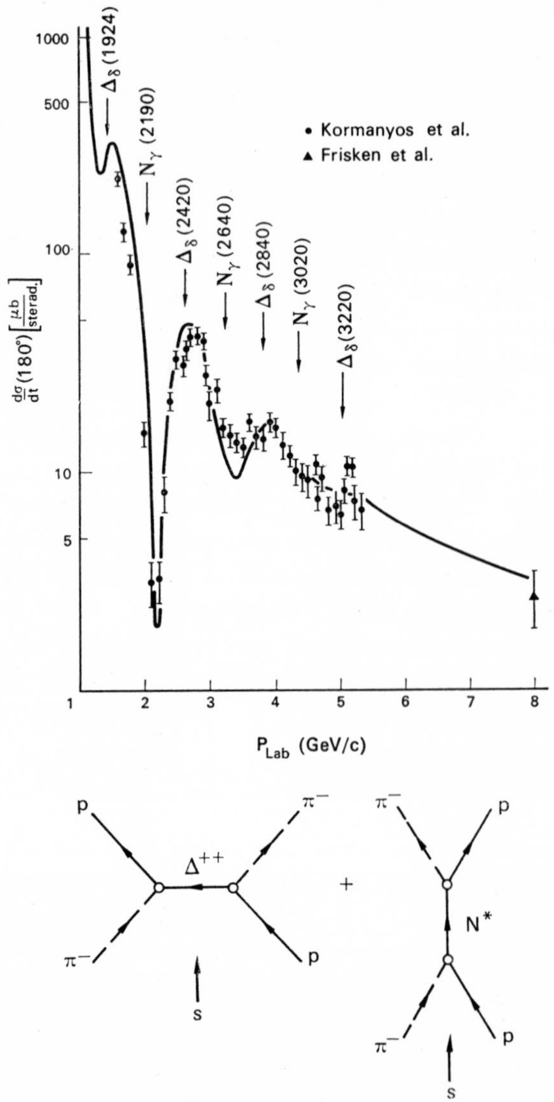


FIG. 7.29. Analysis by Barger and Cline (1966) of backward $\pi^- p$ scattering data as a function of energy.

high energies can be generated by assuming an appropriate string of resonances in the $s$-channel. Alternatively Regge amplitudes in the $t$- or $u$-channels should lead to resonances in the $s$-channel.

Approaches of this type appear to work. Dolen, Horn and Schmid (1968) used the data on low-energy pion charge exchange scattering, $\pi^- p \rightarrow \pi^0 n$, together with the finite energy sum rules to calculate the parameters of the $\varrho$ trajectory which is assumed to dominate the high-energy scattering [§ 7.4(d)]. They obtained a linear trajectory $\alpha(t)$ which had the value $\alpha(t) = 1{\cdot}0 \pm 0{\cdot}3$ at $t = m_\varrho^2$ [compare § 7.4(d)]. A reverse calculation was made by Schmid (1968) with impressive results. In this he

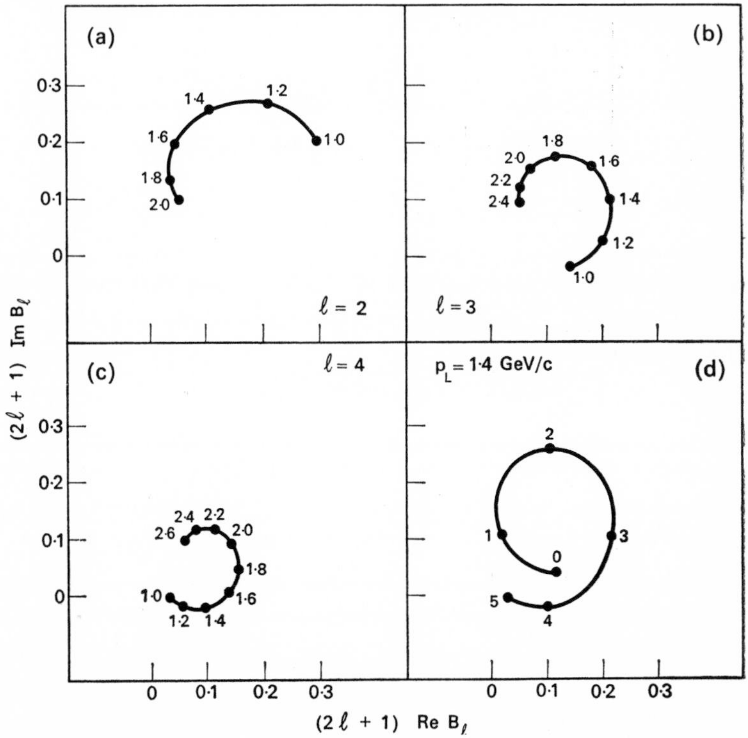


FIG. 7.30. (a)–(c) Behaviour of Im $B_l$ as a function of laboratory momentum (numbers) for fixed $l$. (d) Fixed laboratory momentum, varying $l$.

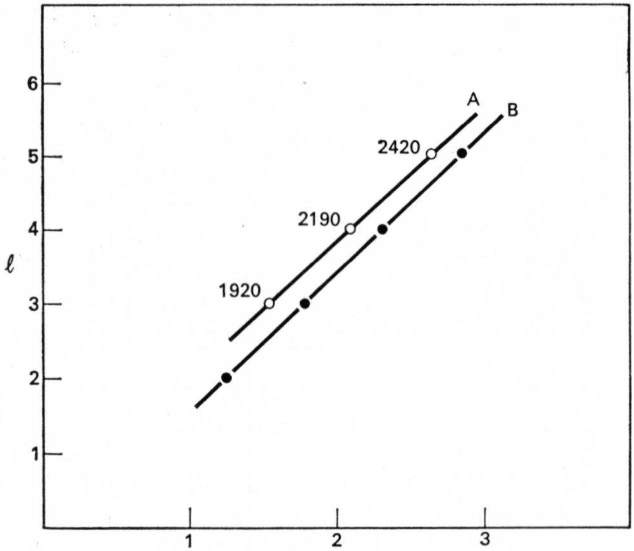


FIG. 7.30 *(cont.)*

made a stronger assumption than in (7.6.3), namely that the equivalence of $T_{\text{Regge}}$ and $T_{\text{Resonance}}$ holds at every point instead of being averaged over a finite region

$$T_{fi}(\nu) \sim T_{\text{Regge}}(\nu) \sim T_{\text{Resonance}}(\nu). \qquad (7.6.4)$$

The reaction studied was again $\pi^- p \rightarrow \pi^0 n$. The Regge amplitude for the $\varrho$-trajectory, $T(\alpha_\varrho)$, was assumed to dominate the high-energy scattering in the $t$-channel and the partial waves in the $s$-channel $B_l(E)$ were extracted with the aid of a projection operator (Appendix A3)

$$B_l(E) = \tfrac{1}{2} \int_{-1}^{1} d(\cos\theta_s)\, T(\alpha_\varrho; E, \cos\theta_s)\, P_l\, (\cos\theta_s).$$

The results for $B_l(E)$ are shown in Fig. 7.30 for $E$ corresponding to a range of laboratory momenta of 1·0 to 2·6 GeV. The curves show the circular, anticlockwise behaviour associated with resonances. The resonance energies were taken from the points corresponding to the tops of the circles and the results shown (line *B*). They are compared with the known $N^*$ resonances (line *A*).

Despite the apparently satisfactory nature of these results, objections have been made that the Regge amplitude cannot be associated with resonances, since the *s*-channel amplitudes obtained by the methods described above do not exhibit poles in the second sheet of the energy plane at $s = m_R^2 - im_R\Gamma$ (Collins *et al.*, 1968). This is a technical requirement for the analytic properties of resonances (Peierls, 1954). At present the conflicting arguments and conclusions associated with the finite energy sum rules, duality and interference models have not been resolved.

## 7.7. The Veneziano Model

The duality concept implies that there is some kind of equivalence in representing an amplitude as a sum of cross-channel $(t, u)$ Regge poles and a sum of direct channel $(s)$ resonances. Nevertheless, it remains to be shown that a finite number of Regge poles can build up a resonance with the correct analytic properties, since none of them have poles in the second sheet of the energy plane. Nor has it been shown that a finite number of *s* channel resonances can lead to a single Regge pole with the asymptotic property $s^\alpha$.

A model has been constructed by Veneziano (1968), however, which explicitly shows how an infinite sum of resonances can combine to give Regge asymptotic behaviour and at the same time the Regge terms yield resonance poles. The model has the further advantages of displaying explicit crossing symmetry, and satisfying the finite energy sum rules.

Veneziano examined possible amplitudes for the process $\pi\pi \to \pi\omega$ which has the nice property of being symmetric in all three channels (Fig. 7.31). The simplest contribution is indicated in the figure. The amplitude can be written as

$$T_{fi} = \varepsilon_{\mu\varrho\sigma\nu}\, e_\mu k_{1\varrho} k_{2\sigma} k_{3\nu} A(s, t, u) \tag{7.7.1}$$

[compare (7.1.28)], and Veneziano chose the following form for *A*

$$A = \frac{\beta}{\pi}\left[\frac{\Gamma(1-\alpha_s)\,\Gamma(1-\alpha_t)}{\Gamma(2-\alpha_s-\alpha_t)} + s \leftrightarrow u + t \leftrightarrow u\right] \tag{7.7.2}$$

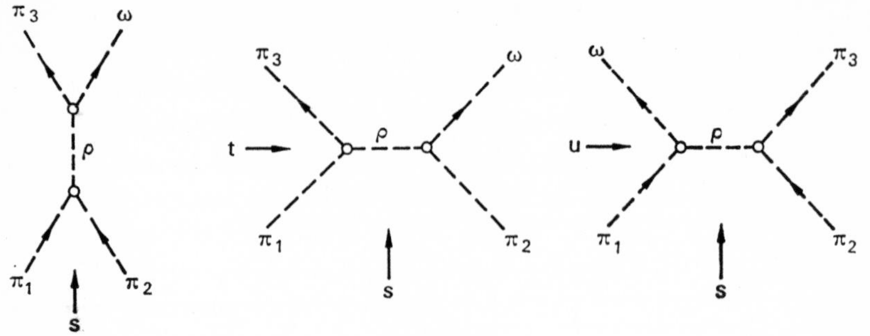


FIG. 7.31.

where $\Gamma$ is the gamma function, $\alpha_s$ the Regge trajectory $\alpha(s)$, etc., and $\beta$ is a normalising constant. The trajectories, are assumed to rise linearly with the appropriate variables $s$, $t$ and $u$

$$\alpha_x = \alpha(0) + \alpha' x \qquad x = s, t, u. \tag{7.7.3}$$

Before considering equation (7.7.2) in detail let us examine the simpler function

$$\frac{\Gamma(-\alpha_s)\,\Gamma(-\alpha_t)}{\Gamma(-\alpha_s - \alpha_t)} \tag{7.7.4}$$

which has the same basic properties. These are as follows:

(1) The $\Gamma$ function possesses the property that $\Gamma(z)$ is an analytic function of $z$ except for poles at the real integers $z = 0, -1, -2. \ldots$ Consequently poles will appear in equation (7.7.4) for $\alpha_s$ or $\alpha_t = 0, 1, .2. \ldots$ No double poles can appear, however, since if both $\alpha_s$ and $\alpha_t$ are non-negative integers, the denominator $\Gamma(-\alpha_s - \alpha_t)$ also has a pole to cancel it.

The behaviour of the function (7.7.4) can therefore be represented schematically as in Fig. 7.32. In this figure the position of the poles in $\alpha_s$ and $\alpha_t$ arising from the numerator are indicated by the horizontal and vertical lines respectively. The denominator $\Gamma(-\alpha_s - \alpha_t)$ leads to zeros in (7.7.4) when $\alpha_s + \alpha_t = 0, 1, 2, \ldots$, and these are indicated by the sloping lines. The functions of the dashed and dotted lines will be discussed later.

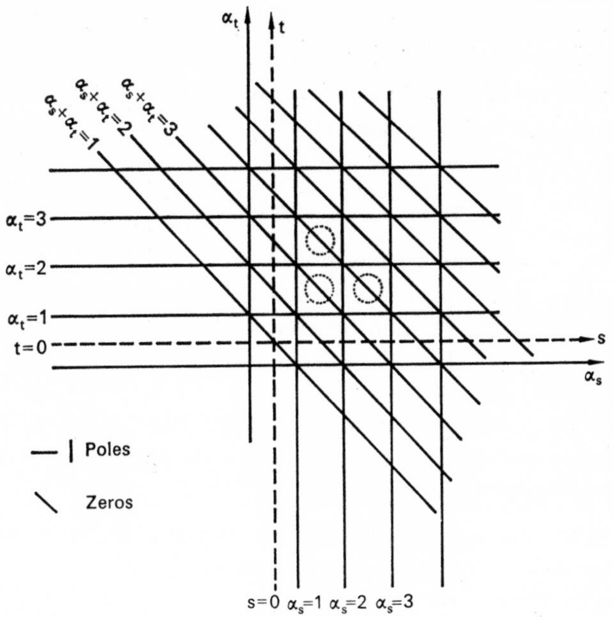


FIG. 7.32.

(2) For purposes of orientation take $\alpha_s = j$ (integer) in (7.7.4), that is a pole in the $s$-channel. Since the gamma function has the property

$$\Gamma(z) = (z - 1)\,\Gamma(z - 1) \tag{7.7.5}$$

repeated applications of this relation shows that the residue is a polynomial of order $j$ in $\alpha_t$

$$\frac{\Gamma(-\alpha_t)}{\Gamma(-\alpha_t - j)} = (-\alpha_t - 1)(-\alpha_t - 2)\cdots(-\alpha_t - j)$$

$$= \sum_{i=0}^{j} C_i \alpha_t^i$$

and since $\alpha_t$ is related to $t$ (7.7.3) and $t = -2k^2(1 - \cos\theta_s)$ the residue is therefore a polynomial of order $j$ in $\cos\theta_s$

$$\frac{\Gamma(-\alpha_t)}{\Gamma(-\alpha_t - j)} = \sum_{i=0}^{j} D_i \cos^i \theta_s.$$

A resonance of spin $j$ involves a Legendre polynomial $P_j(\cos\theta_s)$, and the coefficients $D_l$ in the above expression differ from those in the Legendre polynomial. However, it is possible to develop the summation as a sum of Legendre polynomials

$$\sum_{l=0}^{j} D_l \cos^l \theta_s = \sum_{l=0}^{j} E_l P_l(\cos\theta_s)$$

and so the residue can be interpreted as a family of resonances with spins between 0 and $j$. The systems with spins $0 \rightarrow j-1$ are called daughters. The linear relation between $\alpha$ and $s$ (7.7.3) then implies that the poles in (7.7.4) should represent resonances lying on a series of linearly rising trajectories.

The concept of linearly rising trajectories does seem to fit the experimental data. We showed this for the rho trajectory in Fig. 7.26. In Fig. 7.33 the same point is illustrated for two baryon trajectories—the experimental data is more plentiful for baryon than boson resonances and so the trajectories can be traced more easily.

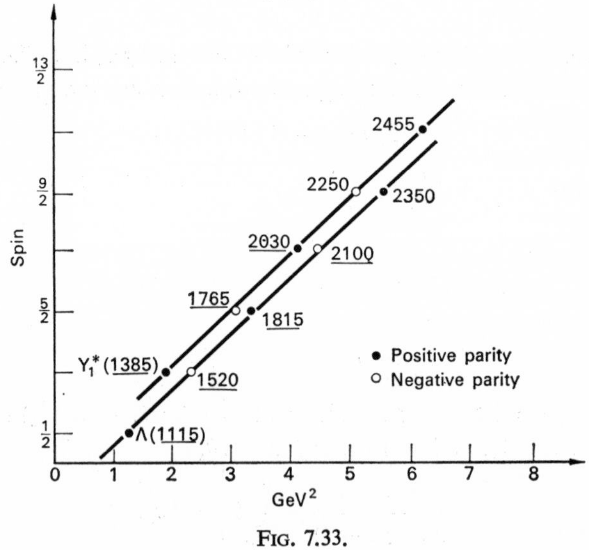


FIG. 7.33.

(3) Consider next the situation when $s \to \infty$, then $\alpha_s \to \alpha' s$ and the gamma function behaves as follows for argument $z \to \infty$

$$\frac{\Gamma(z+a)}{\Gamma(z+b)} \xrightarrow[z\to\infty]{} z^{a-b}. \tag{7.7.6}$$

Equation (7.7.4) then becomes

$$\frac{\Gamma(-\alpha_s)\,\Gamma(-\alpha_t)}{\Gamma(-\alpha_s-\alpha_t)} \xrightarrow[s\to\infty]{} -\frac{\pi}{\Gamma(\alpha_t+1)}\,\frac{1}{\sin\pi\alpha_t}(\alpha' s)^{\alpha_t} \tag{7.7.7}$$

where we have used the following property of the gamma function

$$\Gamma(-z)\,\Gamma(z+1) = \frac{-\pi}{\sin\pi z}.$$

It can be seen that the amplitude (7.7.7) has Regge-like behaviour in the $s$ variable arising from $t$ channel exchanges.

(4) It is apparent from (2) and (3) that the same amplitude can be expressed as a sum of poles in the $s$ or $t$ variables and therefore satisfies duality.

To sum up the Veneziano amplitude displays:

(1) explicit crossing symmetry as shown in (7.7.2),

(2) analyticity,

(3) resonances on linear trajectories,

(4) Regge asymptotic behaviour,

(5) duality.

More detailed considerations show that the amplitude also satisfies the finite energy sum rules. It has certain drawbacks, however; firstly it does not satisfy unitarity and secondly it cannot be applied to particles of half integral spin in a wholly convincing manner.

Let us now return to the original Veneziano formula (7.7.2). Firstly consider (7.7.1), however; the expression $\varepsilon e k_1 k_2 k_3$ took care of the kinematic terms in the amplitude by preserving the symmetry properties appropriate to the $\omega 3\pi$ system (compare (7.1.28)). We are then left with the scalar function $A(s, t, u)$ which is given in detail in (7.7.2). The numbers 1 and 2 in this formula preserve the appropriate symmetries for the internal trajectories in much the same manner as the kinematic terms did for the external lines. The function $\Gamma(1-\alpha_s)$ will lead to poles at $\alpha_s \equiv j = 1, 2, 3, \ldots$ Poles in even values of $j$ would not be expected to appear from invariance arguments. [Since the isospin of the $\omega$ is zero we must exchange isospin one objects according to Fig. 7.31. The $\pi^+\pi^-$ state coupled to this system must then satisfy the relation $(-1)^J = -1$, as shown in our discussion of $\omega$ decay (§ 7.1.(c)). Thus we only expect $j = 1, 3, 5, \ldots$, which is of course appropriate to the rho trajectory.] Even values of $j$ can be made to vanish if we also consider the other terms in the Veneziano formula and apply the following restriction

$$\alpha_s + \alpha_t + \alpha_u = 2. \tag{7.7.8}$$

The plausibility of this restriction follows immediately from the assumption of linearly arising trajectories (7.7.3)

$$\begin{aligned} \alpha_s + \alpha_t + \alpha_u &= 3\alpha_\varrho(0) + \alpha'_\varrho(s + t + u) \\ &= 3\alpha_\varrho(0) + \alpha'_\varrho(3m_\pi^2 + m_\omega^2) \end{aligned}$$

compare (4.3.3); the parameters for the rho trajectory (7.4.14) then give the sum $\sim 2$. If we take the exact relation (7.7.8)

$$2 - \alpha_s - \alpha_u = \alpha_t \qquad 2 - \alpha_t - \alpha_u = \alpha_s$$

$$1 - \alpha_u = \alpha_s + \alpha_t - 1$$

then with a little labour and use of the property

$$\Gamma(z)\,\Gamma(1-z) = \frac{\pi}{\sin \pi z} \tag{7.7.9}$$

the residue at the pole positions in the $s$-channel $\alpha_s \equiv j = 1, 2, 3 \ldots$ becomes

$$\frac{\beta}{\pi}\,\frac{\Gamma(1-\alpha_t)}{\Gamma(2-j-\alpha_t)}\,[1-(-1)^J] \tag{7.7.10}$$

and pole terms only appear for $j = 1, 3, 5, \ldots$ as required by a rho trajectory. Use of the condition (7.7.8) also leads in the high energy limit $s \to \infty$ to the appropriately symmetrised form of the Regge amplitude, but plus an embarrassing string of poles (of zero width) at odd values of

$j$ in the $s$ channel. The latter can be removed by allowing $\alpha_s$ to have an imaginary component

$$\alpha_s = \alpha(0) + \alpha' s + i\sqrt{4m_\pi^2 - s}.$$

This technique moves the pole from the real axis, and yields broad resonances which vanish in the limit $s \to \infty$.

Attempts to apply the Veneziano formula to experimental data have met with partially successful results. The earliest success was that of Lovelace (1968) who examined data on the reaction $\bar{p}n \to 2\pi^-\pi^+$ for antiprotons at rest. The $\bar{p}n$ system at rest has spin-parity $j^P = 0^-$ and the same internal quantum mumbers as the pion; he therefore treated this combination as a massive pion. He thus effectively reduced the problem to that of a system with four external lines (compare Fig. 7.31).

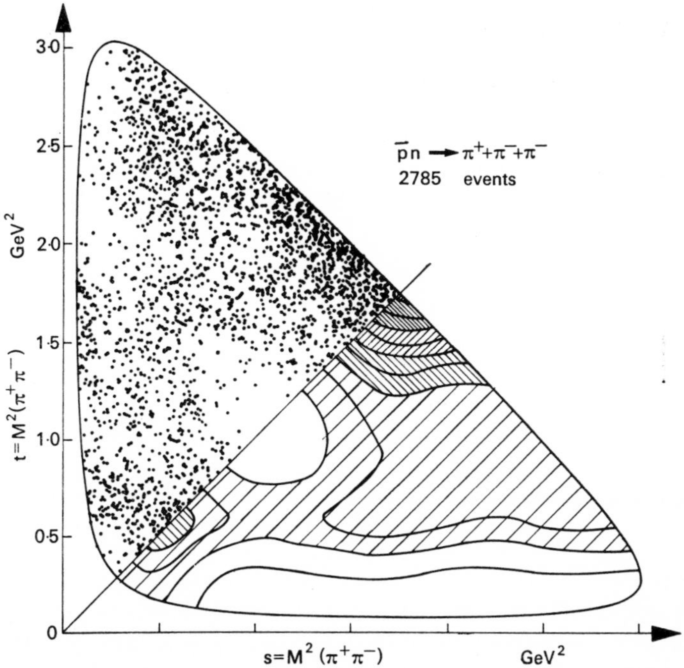


FIG. 7.34. Comparison of experimental (top left) and calculated (bottom right) data for $\bar{p}n \to 2\pi^-\pi^+$ using the Veneziano model (Lovelace, 1968).

In this situation both $s$ and $t$ are positive. A comparison between theory and experiment is shown in Fig. 7.34 and the agreement can be seen to the impressive.

We shall not repeat the details of the calculation, but shall give a qualitative explanation of what is probably the most interesting feature—the pronounced hole in the distribution at $s = t \sim 1$ GeV$^2$. It was pointed earlier that the basic Veneziano amplitude (7.7.4) vanished for $\alpha_s + \alpha_t = 0,1,2,, \ldots$ These zeros were represented by the diagonal lines in Fig. 7.32. The inclusion of imaginary terms in $\alpha$ to yield resonances of finite width effectively causes the zeros to spread to the dotted lines as indicated in Fig. 7.32.

Now the Regge trajectories are linear (7.7.3)

$$\alpha_x = \alpha(0) + \alpha' x \qquad x = s, t, u$$

and for simplicity consider the $\pi\pi$ trajectory as a $\varrho$ trajectory (7.4.14)

$$\alpha_\varrho \sim 0{\cdot}57 + 0{\cdot}91x \qquad x = s, t.$$

We can then sketch in the axes for $s$ and $t$ on Fig. 7.32 (dashed lines) and see that a hole will appear at $s = t \sim 1$ GeV$^2$. (The details are slightly more complicated than we have indicated here.)

Holes have also appeared in data on the process $\bar{p}n \to 2\pi^-\pi^+$ at 1·2 GeV/c (Bettini *et al.*, 1970. I am grateful to Dr. Bettini and his co-workers for providing information prior to publication). Since this experiment was performed at higher incident energy the kinematic regions of $s$ and $t$ are extended. Holes appeared in the plot of $s$ versus $t$ at $s = t = 1$ and 2·17 GeV$^2$, and $s = 1{\cdot}0$, $t = 3{\cdot}2$; $s = 3{\cdot}2$, $t = 1{\cdot}0$ GeV$^2$. A comparison of experimental and theoretical data is displayed in Fig. 7.35. A notable feature of this data is the apparent absence of a pair of holes at $s \sim 2$, $t \sim 1$ and $s \sim 1$, $t \sim 2$ GeV$^2$, which we might have expected from the qualitative argument given above and an inspection of Fig. 7.32. The symmetry requirements (the Veneziano formula was applied assuming the $\bar{p}n$ system to be a $2^+$ meson) partially mask these holes in the theoretical distribution, as an inspection of Fig. 7.35 shows. Nevertheless there do not appear to be any depressions at all in the experimental data.

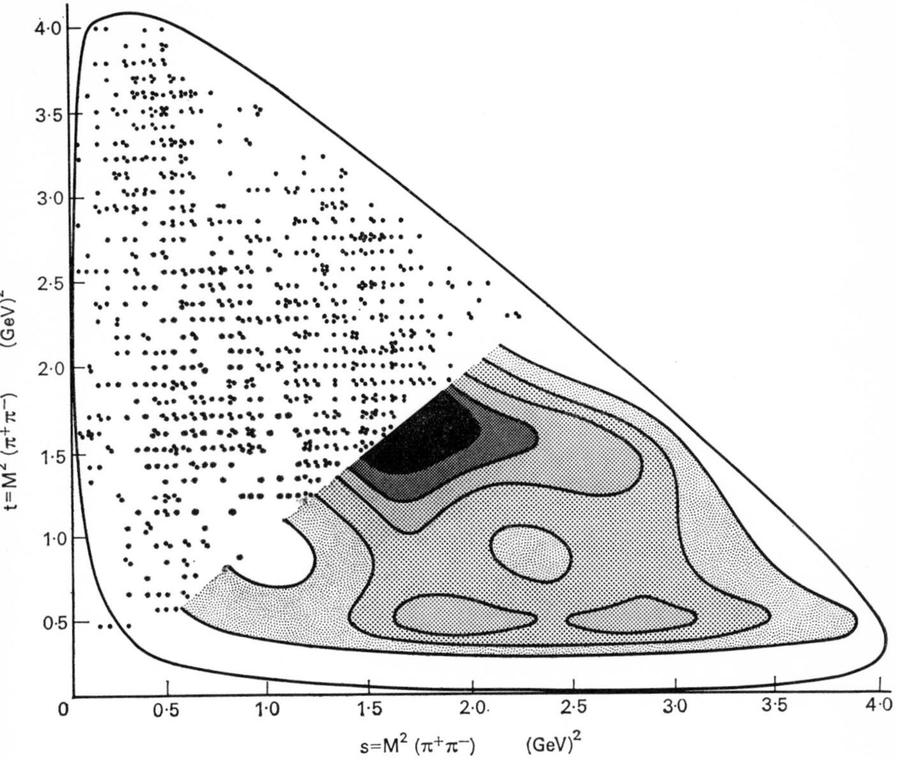


FIG. 7.35.

Interesting extensions of the Veneziano formula to three-body final states have also been made by Petersson and Törnqvist (1969), and Chan Hong-Mo *et al.* (1970). However, it remains to be seen whether a model with the asymptotic properties of a Regge amplitude can ever be successful, when the observed high energy behaviour of many processes does not easily fit into the Regge theory.

# APPENDICES

## A.1. Particle Data

The data given below have been selected from the review of particle properties 1970 (Barbaro-Galtieri *et al.*, 1970).

TABLE 1.

STABLE PARTICLES AND THOSE DECAYING BY WEAK INTERACTIONS

| | $I^G(J^P)\,C$ | Mass (MeV) Mass$^2$ (GeV)$^2$ | Mean life (sec) $c\tau$(cm) | Decays: Partial mode | Decays: Fraction | $p$ or $p_{max}$ (MeV/c) |
|---|---|---|---|---|---|---|
| $\gamma$ | $0, 1(1^-)^-$ | $0(<2{\cdot}0)\ 10^{-21}$ | stable | stable | | |
| $\nu$ | $\nu_e\ J = \frac{1}{2}$ | $0(<60\ \text{eV})$ | stable | stable | | |
| | $\nu_\mu$ | $0(<1{\cdot}6)$ | | | | |
| $e$ | $J = \frac{1}{2}$ | $0{\cdot}511006$ | stable | stable | | |
| | | $\pm 0{\cdot}000002$ | $(> 2 \times 10^{21}\ \text{y})$ | | | |
| $\mu$ | $J = \frac{1}{2}$ | $105{\cdot}659$ | $2{\cdot}1983 \times 10^{-6}$ | $e\nu\nu$ | 100% | 53 |
| | | $\pm 0{\cdot}002$ | $\pm 0{\cdot}0008$ | $e\gamma\gamma$ | $(<1{\cdot}6)\ 10^{-5}$ | 53 |
| | | $m^2 = 0{\cdot}0112$ | $c\tau = 6{\cdot}592 \times 10^4$ | $3e$ | $(<1{\cdot}3)\ 10^{-7}$ | 53 |
| | $m_\mu - m_{\pi^\pm} =$ | $-33{\cdot}920$ | | $e\gamma$ | $(<2)\ 10^{-8}$ | 53 |
| | | $\pm 0{\cdot}013$ | | | | |
| $\pi^\pm$ | $1^-(0^-)$ | $139{\cdot}578$ | $2{\cdot}603 \times 10^{-8}$ | $\mu\nu$ | 100% | 30 |
| | | $\pm 0{\cdot}013$ | $\pm 0{\cdot}006$ | $e\nu$ | $(1{\cdot}24 \pm 0{\cdot}03)\ 10^{-4}$ | 70 |
| | | $m^2 = 0{\cdot}0195$ | $c\tau = 781$ | $\mu\nu\gamma$ | $(1{\cdot}24 \pm 0{\cdot}25)\ 10^{-4}$ | 30 |
| | | $(\tau^+ - \tau^-)/\bar{\tau} =$ | $(0{\cdot}05 \pm 0{\cdot}07)\%$ | $\pi^0 e\nu$ | $(1{\cdot}02 \pm 0{\cdot}07)\ 10^{-8}$ | 5 |
| | | | (test of CPT) | $e\nu\gamma$ | $(3{\cdot}0 \pm 0{\cdot}5)\ 10^{-8}$ | 70 |

TABLE 1 (*cont.*)

| | | | | | | |
|---|---|---|---|---|---|---|
| $\pi^0$ | $1^-(0^-)^+$ | 134·975 | $0{\cdot}89 \times 10^{-16}$ | $\gamma\gamma$ | $(98{\cdot}83 \pm 0{\cdot}04)\%$ | 67 |
| | | | $\pm 0{\cdot}18$ | $\gamma e^+e^-$ | $(1{\cdot}17 \pm 0{\cdot}04)\%$ | 67 |
| | | $m^2 = 0{\cdot}0177$ | $c\tau = 2{\cdot}67 \times 10^{-6}$ | $\gamma\gamma\gamma$ | $(<5)\ 10^{-6}$ | 67 |
| | $m_{\pi^\pm} - m_{\pi^0} =$ | $4{\cdot}6041$ | | $e^+e^-e^+e^-$ | $(3{\cdot}47)\ 10^{-5}$ | 67 |
| | | $\pm 0{\cdot}0037$ | | | | |
| $K^\pm$ | $\frac{1}{2}(0^-)$ | 493·82 | $1{\cdot}235 \times 10^{-8}$ | $\mu\nu$ | $(63{\cdot}77 \pm 0{\cdot}29)\%$ | 236 |
| | | $\pm 0{\cdot}11$ | $\pm 0{\cdot}004$ | $\pi\pi^0$ | $(20{\cdot}93 \pm 0{\cdot}30)\%$ | 205 |
| | | $m^2 = 0{\cdot}244$ | $c\tau = 370$ | $\pi\pi^-\pi^+$ | $(5{\cdot}57 \pm 0{\cdot}04)\%$ | 126 |
| | | $(\tau^+ - \tau^-)/\bar{\tau} =$ | $(0{\cdot}09 \pm 0{\cdot}12)\%$ | $\pi\pi^0\pi^0$ | $(1{\cdot}70 \pm 0{\cdot}05)\%$ | 133 |
| | | | (test of CPT) | $\mu\pi^0\nu$ | $(3{\cdot}18 \pm 0{\cdot}11)\%$ | 215 |
| | | | | $e\pi^0\nu$ | $(4{\cdot}85 \pm 0{\cdot}07)\%$ | 228 |
| | $m_{K^\pm} - m_{K^0} =$ | $-3{\cdot}94$ | | $\pi\pi^\mp e^\pm\nu$ | $(3{\cdot}3 \pm 0{\cdot}3)\ 10^{-5}$ | 203 |
| | | $\pm 0{\cdot}13$ | | $\pi\pi^\pm e^\mp\nu$ | $(<7)\ 10^{-7}$ | 203 |
| | | | | $\pi\pi^\mp\mu^\pm\nu$ | $(0{\cdot}9 \pm 0{\cdot}4)\ 10^{-5}$ | 151 |
| | | | | $\pi\pi^\pm\mu^\mp\nu$ | $(<3)\ 10^{-6}$ | 151 |
| | | | | $e\nu$ | $(1{\cdot}2 \pm 0{\cdot}3)\ 10^{-5}$ | 247 |
| | | | | $\pi\pi^0\gamma$ | $(<1{\cdot}9)\ 10^{-4}$ | 205 |
| | | | | $\pi\pi^+\pi^-\gamma$ | $(10 \pm 4)\ 10^{-5}$ | 126 |
| | | | | $\pi e\nu\gamma$ | $(6 \pm 4)\ 10^{-4}$ | 227 |
| | | | | $\pi e^+e^-$ | $(<0{\cdot}4)\ 10^{-6}$ | 227 |
| | | | | $\pi\mu^+\mu^-$ | $(<2{\cdot}4)\ 10^{-6}$ | 172 |
| | | | | $\pi\gamma\gamma$ | $(<1{\cdot}1)\ 10^{-4}$ | 227 |
| $K^0$ | $\frac{1}{2}(0^-)$ | 497·76 | 50% $K_{\text{short}}$, 50% $K_{\text{long}}$ | | | |
| | | $\pm 0{\cdot}16$ | | | | |

TABLE 1 (*cont.*)

| | $I^G(J^P)C$ | Mass (MeV) Mass² (GeV)² | Mean life (sec) $c\tau$(cm) | Decays: Partial mode | Decays: Fraction | $p$ or $p_{max}$ (MeV/c) |
|---|---|---|---|---|---|---|
| $K_S^0$ | $\frac{1}{2}(0^-)$ | 497.76 | $0{\cdot}862 \times 10^{-10}$ | $\pi^+\pi^-$ | (68·7 )% | 206 |
| | | 0.16 | $\pm 0{\cdot}006$ | $\pi^0\pi^0$ | (31·3 ± 0·6)% | 209 |
| | | $m^2 = 0{\cdot}248$ | $c\tau = 2{\cdot}59$ | $\mu^+\mu^-$ | $(<3{\cdot}1)\ 10^{-7}$ | 225 |
| | | | | $e^+e^-$ | $(<2{\cdot}2)\ 10^{-7}$ | 249 |
| | | | | $\pi^+\pi^-\gamma$ | $(3{\cdot}3 \pm 1{\cdot}2)\ 10^{-3}$ | 206 |
| $K_L^0$ | $\frac{1}{2}(0^-)$ | | $5{\cdot}38 \times 10^{-8}$ | $\pi^0\pi^0\pi^0$ | (21·5 ± 0·7)% | 139 |
| | | | $\pm 0{\cdot}19$ | $\pi^+\pi^-\pi^0$ | (12·6 ± 0·3)% | 133 |
| | | | $c\tau = 1614$ | $\pi\mu\nu$ | (26.8 ± 0·7)% | 216 |
| | | | | $\pi e\nu$ | (38·8 ± 0·8)% | 229 |
| | $m_{K_S} - m_{K_L} = -\ 0{\cdot}469 \times 1/\tau_S$ $\pm 0{\cdot}015$ | | | $\pi^+\pi^-$ | (0·157 ± 0·005)% | 206 |
| | | | | $\pi^0\pi^0$ | (0·121 ± 0·029)% | 209 |
| | | | | $\pi^+\pi^-\gamma$ | $(<0{\cdot}4)\ 10^{-3}$ | 206 |
| | | | $\dfrac{\Gamma(K_S \to \pi^+\pi^-\pi^0)}{\Gamma(K_L \to \pi^+\pi^-\pi^0)} < 0{\cdot}45$ | $\gamma\gamma$ | $(5{\cdot}2 \pm 0{\cdot}5)\ 10^{-4}$ | 249 |
| | | | (test of CP) | $e\mu$ | $(<0{\cdot}6)\ 10^{-5}$ | 238 |
| | | | | $\mu^+\mu^-$ | $(<1{\cdot}5)\ 10^{-6}$ | 225 |
| | | | | $e^+e^-$ | $(<1{\cdot}7)\ 10^{-5}$ | 249 |

TABLE 1 (*cont.*)

| | | | | | | |
|---|---|---|---|---|---|---|
| $\eta$ | $0^+(0^-)^+$ | 548·8 | $\Gamma = (2·63 \pm 0·64)$ keV | $\gamma\gamma$ | $(38·2 \pm 2·1)\%$ | 274 |
| | | $\pm 0·6$ | Neutral decays | $\pi^0\gamma\gamma$ | $(2·0 \pm 2·8)\%$ | 258 |
| | | $m^2 = 0·301$ | 71·5% | $3\pi$ | $(31·4 \pm 2·7)\%$ | 179 |
| | | | | $\pi^+\pi^-\pi^0$ | $(23·0 \pm 1·1)\%$ | 174 |
| | | | Charged decays | $\pi^+\pi^-\gamma$ | $(5·4 \pm 0·5)\%$ | 236 |
| | | | 28·5% | $\pi^0e^+e^-$ | $(<0·01)\%$ | 258 |
| | | | | $\pi^+\pi^-e^+e^-$ | $(0·1 \pm 0·1)\%$ | 236 |
| $p$ | $\frac{1}{2}(\frac{1}{2}^+)$ | 938·256 | stable | | | |
| | | $\pm 0·005$ | $(>2 \times 10^{28}$ y) | | | |
| | | $m^2 = 0·880$ | | | | |
| $n$ | $\frac{1}{2}(\frac{1}{2}^+)$ | 939·550 | $(0·932 \pm 0·014)\ 10^3$ | $pe^-\nu$ | 100% | 1 |
| | | $\pm 0·005$ | $c\tau = 2·80 \times 10^{13}$ | | | |
| | | $m^2 = 0·882$ | | | | |
| | $m_p - m_n =$ | $-1·2933$ | | | | |
| | | $\pm 0·0001$ | | | | |
| $\Lambda$ | $0(\frac{1}{2}^+)$ | 1115·60 | $2·51 \times 10^{-10}$ | $p\pi^-$ | $(65·3 \pm 1·3)\%$ | 100 |
| | | $\pm 0·08$ | $\pm 0·03$ | $n\pi^0$ | $(34·7)\%$ | 104 |
| | | | $c\tau = 7·54$ | $pe\nu$ | $(0·85 \pm 0·07)\ 10^{-3}$ | 163 |
| | | $m^2 = 1·245$ | | $p\mu\nu$ | $(1·35 \pm 0·60)\ 10^{-4}$ | 131 |
| $\Sigma^+$ | $1(\frac{1}{2}^+)$ | 1189·40 | $0·802 \times 10^{-10}$ | $p\pi^0$ | $(51·7)\%$ | 189 |
| | | $\pm 0·19$ | $\pm 0·007$ | $n\pi^+$ | $(48·3 \pm 0·8)\%$ | 185 |
| | | | $c\tau = 2·41$ | $p\gamma$ | $(1·16 \pm 0·17)\ 10^{-3}$ | 225 |
| | | $m^2 = 1·412$ | | $n\pi^+\gamma$ | $(1·3 \pm 0·3)\ 10^{-4}$ | 185 |
| | $m_{\Sigma^+} - m_{\Sigma^-} =$ | $-7·92$ | | $\Lambda e^+\nu$ | $(2·02 \pm ·047)\ 10^{-5}$ | 72 |
| | | $\pm 0·13$ | $\frac{\Gamma(\Sigma^+ \to l^+n\nu)}{\Gamma(\Sigma^- \to l^-n\nu)} = <0·03 \leftarrow$ | $n\mu^+\nu$ | $(<1·1)\ 10^{-5}$ | 202 |
| | | | | $ne^+\nu$ | $(<0·7)\ 10^{-5}$ | 224 |

TABLE 1 (*cont.*)

| | $I^G(J^P)C$ | Mass (MeV) Mass$^2$ (GeV)$^2$ | Mean life (sec) $c\tau$(cm) | Decays: Partial mode | Decays: Fraction | $p$ or $p_{max}$ (MeV/c) |
|---|---|---|---|---|---|---|
| $\Sigma^0$ | $1(\frac{1}{2}^+)$ | 1192·46 | $<1{\cdot}0 \times 10^{-14}$ | $\Lambda\gamma$ | 100% | 75 |
| | | $\pm 0{\cdot}12$ | $c\tau < 3 \times 10^{-4}$ | $\Lambda e^+e^-$ | $(5{\cdot}45)\ 10^{-3}$ | |
| | | $m^2 = 1{\cdot}422$ | | | | |
| $\Sigma^-$ | $1(\frac{1}{2}^+)$ | 1197·32 | $1{\cdot}49 \times 10^{-10}$ | $n\pi^-$ | 100% | 193 |
| | | $\pm 0{\cdot}11$ | $\pm 0{\cdot}03$ | $ne^-\nu$ | $(1{\cdot}06 \pm 0{\cdot}05)\ 10^{-3}$ | 230 |
| | | | $c\tau = 4{\cdot}47$ | $n\mu^-\nu$ | $(0{\cdot}45 \pm 0{\cdot}04)\ 10^{-3}$ | 210 |
| | | $m^2 = 1{\cdot}434$ | | $\Lambda e^-\nu$ | $(0{\cdot}60 \pm 0{\cdot}06)\ 10^{-4}$ | 79 |
| | $m_{\Sigma^0} - m_{\Sigma^-}$ | $= 4{\cdot}86$ | | $n\pi^-\gamma$ | $(1{\cdot}0 \pm 0{\cdot}2)\ 10^{-4}$ | 193 |
| | | $\pm 0{\cdot}07$ | | | | |
| $\Xi^0$ | $\frac{1}{2}(\frac{1}{2}^+)$ | 1314·7 | $3{\cdot}03 \times 10^{-10}$ | $\Lambda\pi^0$ | 100% | 135 |
| | | $\pm 0{\cdot}7$ | $\pm 0{\cdot}18$ | $p\pi^-$ | $(<0{\cdot}9)\ 10^{-3}$ | 299 |
| | | $m^2 = 1{\cdot}728$ | $c\tau = 9{\cdot}10$ | $pe^-\nu$ | $(<1{\cdot}3)\ 10^{-3}$ | 323 |
| | $m_{\Xi^0} - m_{\Xi^-}$ | $= 6{\cdot}6$ | | $\Sigma^+e^-\nu$ | $(<1{\cdot}5)\ 10^{-3}$ | 119 |
| | | $\pm 0{\cdot}7$ | | $\Sigma^-e^+\nu$ | $(<1{\cdot}5)\ 10^{-3}$ | 112 |
| | | | | $\Sigma^+\mu^-\nu$ | $(<1{\cdot}5)\ 10^{-3}$ | 64 |
| | | | | $\Sigma^-\mu^+\nu$ | $(<1{\cdot}5)\ 10^{-3}$ | 49 |
| | | | | $p\mu^-\nu$ | $(<1{\cdot}3)\ 10^{-3}$ | 309 |

TABLE 1 (*cont.*)

| | | | | | | |
|---|---|---|---|---|---|---|
| $\Xi^-$ | $\frac{1}{2}(\frac{1}{2}^+)$ | 1321·25 | $1{\cdot}66 \times 10^{-10}$ | $\Lambda\pi^-$ | 100% | 139 |
| | | ±0·18 | ±0·04 | $\Lambda e^-\nu$ | $(0{\cdot}67 \pm 0{\cdot}23)\ 10^{-3}$ | 190 |
| | | $m^2 = 1{\cdot}746$ | $c\tau = 4{\cdot}98$ | $\Sigma^0 e^-\nu$ | $(<0{\cdot}5)\ 10^{-3}$ | 122 |
| | | | | $\Lambda\mu^-\nu$ | $(<1{\cdot}3)\ 10^{-3}$ | 163 |
| | | | | $\Sigma^0\mu^-\nu$ | $(<0{\cdot}5)$ % | 70 |
| | | | | $n\pi^-$ | $(<1{\cdot}1)\ 10^{-3}$ | 303 |
| | | | | $ne^-\nu$ | $(<1{\cdot}0)$% | 327 |
| $\Omega^-$ | $0(\frac{3}{2}^+)$ | $1672{\cdot}5 \pm 5$ | $1{\cdot}3\ {}^{+0{\cdot}4}_{-0{\cdot}3} \times 10^{-10}$ | $\Xi^0\pi^-$ | Total of | 293 |
| | | $m^2 = 2{\cdot}797$ | | $\Xi^-\pi^0$ | 28 events | 289 |
| | | | $c\tau = 3{\cdot}9$ | $\Lambda K^-$ | seen | 210 |

TABLE 2. MESONS

| Name | $I^G(J^P)\,C_n$ —estab. ? = guess | Mass $M$ (MeV) | Width $\Gamma$ (MeV) | $M^2$ $\pm\Gamma M$ (GeV)$^2$ | Partial decay modes: Mode | Fraction % | $p$ or $p_{max}$ (MeV/c) |
|---|---|---|---|---|---|---|---|
| $\pi^{\pm}(140)$ | $\underline{1^-(0^-)+}$ | 139·58 | 0·0 | 0·019483 | See Table 1 | | |
| $\pi^0(135)$ | | 134·97 | 7·2 eV | 0·018217 | | | |
| | | | ±1·2 eV | | | | |
| $\eta(549)$ | $\underline{0^+(0^-)+}$ | 548·8 | 2·63 keV | 0·301 | All neutral | 71 } See Table 1 | |
| | | ±0·6 | ±0·64 keV | ±0·000 | $\pi^+\pi^-\pi^0+\pi^+\pi^-\gamma$ | 29 } | |
| $\eta_{0^+}(700)$ | $\underline{0^+(0^+)+}$ | ~700 | ≫100 | ~0·5 | $\pi\pi$ | 100 | ~320 |
| "$\varepsilon$" → $\pi\pi$ | $\delta_{00}$ seems to stay near 90° from 650 to 900 MeV | | | | | | |
| $\varrho(765)$ | $\underline{1^+(1^-)-}$ | 765 | 125 | 0·585 | $\pi\pi$ | ~100 | 356 |
| | | ±10 (c) | ±20 (c) | ±0·095 | $\pi^{\pm}\pi^+\pi^-\pi^0$ | <0·2 | 243 |
| | | | | | $\pi^+\pi^-\pi^+\pi^-$ | <0·15 | 243 |
| | | | | | $\pi^{\pm}\gamma$ | <0·2 | 370 |
| | | | | | $\eta\pi^{\pm}$ | <0·8 | 141 |
| | | | | | $e^+e^-$ | 0·0060 ± 0·0006 | 382 |
| | | | | | $\mu^+\mu^-$ | 0·0062 ± 0·0011 | 368 |
| $\omega(784)$ | $\underline{0^-(1^-)-}$ | 783·7 | 12·7 | 0·614 | $\pi^+\pi^-\pi^0$ | 87 ± 4 | 328 |
| | | ±0·4 | ±1·2 | ±0·100 | $\pi^+\pi^-$ | >0·3 (95% confidence) | 366 |
| | | | | | $\pi^0\gamma$ | 9·4 ± 1·7 | 380 |
| | | | | | $e^+e^-$ | 0·0066 ± 0·0017 | 392 |

TABLE 2 (*cont.*)

| | | | | | | | |
|---|---|---|---|---|---|---|---|
| $\eta'(958)$ or $X^0$ | $\underline{0}^+(0^-)\,\underline{+}$ | 957·7 ±0·8 | <4 | 0·917 <0·004 | $\eta\pi\pi$ | 66 ± 4 | 231 |
| | | | | | $\varrho^0\gamma$ | 30 ± 3 | 173 |
| | $J^P = 2^-$ not excluded | | | | $\gamma\gamma$ | 4·7 ± 2·9 | 479 |
| $\delta(962)$ | ≧1( ) | 962 ±5 | <5 | 0·927 <0·005 | $\eta\pi$ possibly seen | | 305 |
| | These two could be related | | | | | | |
| $\pi_N(1016) \to K\bar{K}$ | $\underline{1}^-(0^+)$ + | 1016 ±10 | ≈25 if res. | 1·032 ±0·025 | $K^{\pm}K^0$ | Only mode seen | 111 |
| | | | | | $\eta\pi$ | <80 | 342 |
| Resonance, virtual bound state, or antibound state, still not distinguished | | | | | | | |
| $\phi(1019)$ | $\underline{0^-(1^-)\,-}$ | 1019·5 ±0·6 | 3·9 ±0·4 | 1·039 ±0·004 | $K^+K^-$ | 45·5 ± 3·3 | 126 |
| | | | | | $K_LK_S$ | 36·4 ± 3·4 | 110 |
| | | | | | $\pi^+\pi^-\pi^0$ (incl. $\varrho\pi$) | 18·1 ± 4·9 | 462 |
| | | | | | $e^+e^-$ | 0·036 ± 0·003 | 510 |
| | | | | | $\mu^+\mu^-$ | $0{\cdot}035^{+0{\cdot}035}_{-0{\cdot}018}$ | 499 |
| $\eta_{0^+}(1060)$ "S*" $\to K_SK_S$ | $\underline{0^+(1^+)\,+}$ if res. | 1062 ±20 | ~80(?) | 1·13 ±0·09 | $\pi\pi$ | <65 | 513 |
| | | | | | $K\bar{K}$ | >35 | 190 |
| Resonance and scattering length both possible | | | | | | | |
| $A1(1070)$ | $\underline{1}^-(1^+)\,\underline{+}$ | 1070 ±20 | 95 ±35 | 1·14 ±0·10 | $3\pi$ | ~100 | 488 |
| | | | | | $K\bar{K}$ | <0·25 | 201 |
| Interpretation still slightly in doubt; $J^P = 2^-$ not excluded | | | | | $[G = (-1)^{l+I}$ forbids $K\bar{K}]$ | | |

TABLE 2 (*cont.*)

| Name | $I^G(J^P)\ C_n$ —estab. ? = guess | Mass $M$ (MeV) | Width $\Gamma$ (MeV) | $M^2$ $\pm\Gamma M$ (GeV)$^2$ | Partial decay modes: Mode | Fraction % | $p$ or $p_{max}$ (MeV/c) |
|---|---|---|---|---|---|---|---|
| $B$(1235) | $\underline{1}^+(1^+)\ \underline{-}$ | 1235 | *102* | 1·53 | $\omega\pi$ | ~100 | 350 |
| | | ±15 | ±20 | ±0·13 | $\pi\pi$ | <30} Absence suggests | 602 |
| | | | | | $K\bar{K}$ | < 2} $J^P$ = Abn. | 371 |
| $f$(1260) | $\underline{0^+(2^+)\ +}$ | 1264 | 151 | 1·60 | $\pi\pi$ | ~100 | 616 |
| | | ±10 | ±25 | ±0·19 | $2\pi^+2\pi^-$ | < 4 | 553 |
| | | | | | $K\bar{K}$ indic. seen | ~ 3 | 389 |
| $D$(1285) | $\underline{0}^+(A)\ +$ | 1288 | 33 | 1·66 | $K\bar{K}\pi$ [mainly $\pi_N(1016)\ \pi$] | Seen | 307 |
| | | ±7 | ±5 | ±0·04 | $\pi\pi\eta$ | Possibly large | 485 |
| | | | | | $\pi\pi\varrho$ | Not seen | 354 |
| | $J^P = 0^-, 1^+, 2^-$, with $1^+$ favoured | | | | | | |
| $A2_L$(1280) | $1^-(2^+)\ +$ | 1280 | 22 | 1·64 | $\varrho\pi$ (and $\pi$ + neutrals) | Dominant | 395 |
| | | ±4 | ±4 | ±0·028 | $K\bar{K}$ | Seen | 405 |
| | | | | | $\eta\pi$ | Indication seen | 511 |
| $A2_H$(1320) | $1^-(2^+)\ +$ | 1320 | 21 | 1·74 | $\varrho\pi$ (and $\pi$ + neutrals) | Dominant | 423 |
| | | ±5 | ±4 | ±0·028 | $K\bar{K}$ | Seen | 436 |
| | | | | | $\eta\pi$ | Indication seen | 535 |

TABLE 2 (*cont.*)

| | | | | | | | |
|---|---|---|---|---|---|---|---|
| $E(1422)$ | $0^+(0^-)+$ | 1422 | 69 | 2·02 | $K^*\bar{K}+\bar{K}^*K$ | $50 \pm 10$ } (so 100% | 153 |
| | | $\pm 4$ | $\pm 8$ | $\pm 0{\cdot}10$ | $\pi_N(1016)\,\pi$ | $50 \pm 10$ } $\bar{K}K\pi$) | 326 |
| | $J^P = 1^+$ not excluded | | | | $\pi\pi\eta$ | $<60$ | 568 |
| | | | | | $\pi\pi\varrho$ | Not seen | 457 |
| $f'(1514)$ | $0^+(2^+)+$ | 1514 | 73 | 2·29 | $K\bar{K}$ | $72 \pm 12$ | 570 |
| | | $\pm 5$ | $\pm 23$ | $\pm 0{\cdot}11$ | $K^*\bar{K}+\bar{K}^*K$ | $10 \pm 10$ | 294 |
| | | | | | $\pi\pi$ | $<14$ | 744 |
| | | | | | $\eta\pi\pi$ | $18 \pm 10$ | 624 |
| | | | | | $\eta\eta$ | $<40$ | 521 |
| $\pi/\varrho(1540)$ | 1 (A) | 1540 | 40 | 2·37 | $K^*\bar{K}+\bar{K}^*K$ | Only mode seen | 321 |
| "$F_1$" $\to K^*\bar{K}$ | $-$ ? | $\pm 5$ | $\pm 15$ | $\pm 0{\cdot}06$ | | | |
| | Seen in only one experiment | | | | | | |
| $\pi_A(1640)$ | $1^-(A)+$ | 1633 | 93 | 2·67 | $3\pi$ | Dominant | 788 |
| $\to 3\pi$ | | $\pm 9$ | $\pm 24$ | $\pm 0{\cdot}15$ | $[\pi^{\pm}\varrho^0/\text{all}\,\pi^+\pi^+\pi^-$ | $<40]$ | 629 |
| | | | | | $[\pi^{\pm}f(\to\pi^+\pi^-)/\text{all}\,\pi^{\pm}\pi^+\pi^-$ | $35\,^{+25}_{-20}]$ | 304 |
| | $J^P = 2^-$ preferred | | | | $\omega\pi\pi$ | Possibly observed | 592 |
| | | | | | $\omega\varrho$ | Possibly observed | 259 |
| | Low signal/background; may be partly | | | | $\pi\eta$ | $<9$ | 717 |
| | Deck effect and (or) several resonances | | | | $K\bar{K}$ | Not seen | 647 |
| | | | | | $\pi^{\pm}2\pi^+2\pi^-$ | $<10$ | 710 |

—[$R1(1630)$ is included in both the $\pi_A(1640)$ and $\varrho_N(1660)$ listings]

TABLE 2 (*cont.*)

| Name | $I^G(J^P)\,C_n$ estab. ? = guess | Mass $M$ (MeV) | Width $\Gamma$ (MeV) | $M^2$ $\pm \Gamma M$ (GeV)$^2$ | Partial decay modes: Mode | Fraction % | $p$ or $p_{max}$ (MeV/c) |
|---|---|---|---|---|---|---|---|
| $\varrho_N(1660)$ | $\underline{1^+(N)}\,\underline{-}$ | 1663[p] | 111 | 2·77 | $2\pi$ | Dominant | 820 |
| "$g$" $\to 2\pi$ | | ±20 | ±30 | ±0·18 | $K\bar{K}$ | $8^{+8}_{-3}$ | 666 |
| | ($J^P = 1^-, 3^-, \ldots$ with $3^-$ favoured) | | | | (Other modes under $\varrho(1710)$) | | |
| $\varrho(1710)$? | $\underline{1^+(\ )}\,\underline{-}$ | 1714 | 110 | 2·94 | $4\pi$ | Dominant | 799 |
| $\to 4\pi$ | | ±20 | ±25 | ±0·08 | $[\pi^\pm A_2^0(\to\pi^+\pi^-\pi^0)/\text{all}\,\pi^\pm\pi^+\pi^-\pi^0$ | 40 ± 20] | 342 |
| | Not yet clear whether this is just an alternative mode of the $\varrho_N(1660)$, or a different resonance; the branching ratios are therefore only tentative | | | | $[\pi^\pm\omega(\to\pi^+\pi^-\pi^0)/\text{all}\,\pi^\pm\pi^+\pi^-\pi^0$ | 25 ± 10] | 669 |
| | | | | | $[\varrho^\pm\varrho^0$ | Seen] | 386 |
| | | | | | $\pi^\pm\phi/\pi^\pm\pi^+\pi^-\pi^0$ | < 11 | 542 |
| | | | | | $\pi^\pm 2\pi^+ 2\pi^-\pi^0$ | < 15 | 705 |
| | | | | | $2\pi(\text{if} \neq \varrho_N(1660))$ | < 10 | 846 |
| Also bumps grouped as $R(1750)$, $S(1930)$, $\varrho(2100)$, $T(2200)$, $\varrho(2275)$, and $N\bar{N}(2345)$. | | | | | | | |
| $U(2375)$ | $1^+(\ )\ -$ | 2371 | 30 | 5·62 | Seen in $\pi^- p \to pU^-$ and | | |
| | | ±8 | ±20 | ±0·07 | $p\bar{p} \to K_1^0 K_1^0 \omega, K_1 K_2(n\pi^0)$ | | |
| Also 5 bumps, $N\bar{N}(2380)$, $X^-(2500)$, $X^-(2620)$, $X^-(2800)$, $X^-(2880)$. | | | | | | | |
| $K^+(494)$ | $\underline{1/2(0^-)}$ | 493·82 | | 0·244 | See Table 1 | | |
| $K^0(498)$ | | 497·76 | | 0·248 | | | |

TABLE 2 (*cont.*)

| | | | | | | | |
|---|---|---|---|---|---|---|---|
| | | Charged $K^*$ | | | | | |
| $K^*(892)$ | 1/2(1⁻) | (±)892.1<br>±0·4<br>($m_0 - m_\pm = 7 \pm 3$) | 50.1<br>±0·8 | 0·796<br>±0·045 | $K\pi$<br>$K\pi\pi$ | ~ 100<br>0·2 | 288<br>216 |
| $K_A(1240)$ or $C$ | 1/2(1⁺) | 1243<br>±6 | 90<br>±40 | 1·54<br>±0·11 | $Q$ region | $K\pi\pi$ Only mode seen<br>[$K^*\pi$ Large]<br>[$K\varrho$ Seen] | 478<br>276<br>0 |
| $K_A$(1280 to 1360) ? | 1/2(1⁺) | 1280 to 1360 | | | | | |
| | $J^P = 2^-$ not completely ruled out | | | | | | |
| $K_N(1420)$ | 1/2(2⁺) | (±)1409<br>±4 | 96<br>±7 | 1·985<br>±0·135 | $K\pi$<br>$K^*\pi$<br>$K\varrho$<br>$K\omega$<br>$K\eta$ | 49·2 ± 3·4<br>36·3 ± 3·1<br>8·0 ± 3·5<br>4·2 ± 1·3<br>2·2 ± 1·6 | 609<br>406<br>311<br>291<br>474 |
| | $J^P = 3^-$ still possible | | | | | | |
| $K_A(1775)$ or $L$ | 1/2($A$) | 1775 | | 3·15 | $K\pi\pi$<br>[$K^*(1420)\,\pi$ | Only mode seen<br>Large] | 794<br>305 |
| Interpretation in doubt | $J^P = 1^+, 2^-$ favoured | | | | | | |

TABLE 3a. BARYONS $S = 0$

| | Particle or resonance | $I(J^P)$ | $\pi$ or $K$ beam: $T$ (GeV), $p$ (GeV/c), $\sigma = 4\pi\lambda\!\!\!^{-2}$ (mb) | Mass (MeV) | $\Gamma$ (MeV) | $M^2 \pm \Gamma M$ (GeV$^2$) | Decay modes: Partial mode | Decay modes: Fraction % | $p$ or $p_{max}$ (MeV/c) |
|---|---|---|---|---|---|---|---|---|---|
| | $p$<br>$n$ | 1/2(1/2$^+$) | | 938·3<br>939·6 | | 0·880<br>0·883 | see Table 1 | | |
| | $N'(1470)$ | 1/2(1/2$^+$) $P'_{11}$ | $T = 0{\cdot}53\pi p$<br>$p = 0{\cdot}66$<br>$\sigma = 27{\cdot}8$ | 1435 to 1505 | 200 to 400 | 2·16 ±0·36 | $N\pi$<br>$N\pi\pi$ | 60<br>40 | 420<br>368 |
| | $N'(1520)$ | 1/2(3/2$^-$) $D'_{13}$ | $T = 0{\cdot}61$<br>$p = 0{\cdot}74$<br>$\sigma = 23{\cdot}5$ | 1510 to 1540 | 105 to 150 | 2·31 ±0·18 | $N\pi$<br>$N\pi\pi$<br>$N\eta$ | 50<br>50<br>~0·5 | 456<br>410<br>150 |
| $N$ | $N'(1535)$ | 1/2(1/2$^-$) $S'_{11}$ | $T = 0{\cdot}64$<br>$p = 0{\cdot}76$<br>$\sigma = 22{\cdot}5$ | 1500 to 1600 | 50 to 160 | 2·36 ±0·18 | $N\pi$<br>$N\eta$<br>$N\pi\pi$ | 34<br>66<br>small | 467<br>182<br>422 |
| | $N(1670)$ | 1/2(5/2$^-$) $D_{15}$ | $T = 0{\cdot}87$<br>$p = 1{\cdot}00$<br>$\sigma = 15{\cdot}6$ | 1655 to 1680 | 105 to 175 | 2·79 ±0·24 | $N\pi$<br>$N\pi\pi$<br>$\Lambda K$<br>$N\eta$ | 42<br>58<br><0·1<br><0·3 | 560<br>525<br>200<br>368 |
| | $N(1688)$ | 1/2(5/2$^+$) $F_{15}$ | $T = 0{\cdot}90$<br>$p = 1{\cdot}03$<br>$\sigma = 14{\cdot}9$ | 1680 to 1692 | 105 to 180 | 2·85 ±0·21 | $N\pi$<br>$N\pi\pi$<br>$\Lambda K$<br>$N\eta$ | 60<br>40<br><0·1<br><0·2 | 572<br>538<br>231<br>388 |

TABLE 3a (*cont.*)

| | | | | | | | | | |
|---|---|---|---|---|---|---|---|---|---|
| N | $N''(1700)$ | 1/2(1/2⁻) $S''_{11}$ | $T = 0{\cdot}92$ | 1665 to | 100 to | 2·89 | $N\pi$ | 70 | 580 |
| | | | $p = 1{\cdot}05$ | 1765 | 400 | ±0·44 | $\Lambda K$ | 5 | 250 |
| | | | $\sigma = 14.3$ | | | | $N\eta$ | ~3 | 340 |
| | $N''(1780)$ | 1/2(1/2⁺) $P''_{11}$ | $T = 1{\cdot}07$ | 1750 to | 270 to | 3·17 | $N\pi$ | 34 | 633 |
| | | | $p = 1{\cdot}20$ | 1860 | 450 | ±0·62 | $\Lambda K$ | ~1 | 353 |
| | | | $\sigma = 12{\cdot}2$ | | | | $N\eta$ | ~10 | 476 |
| | $N(1860)$ | 1/2(3/2⁺) $P_{13}$ | $T = 1{\cdot}22$ | 1840 to | 310 to | 3·46 | $N\pi$ | 27 | 685 |
| | | | $p = 1{\cdot}36$ | 1900 | 450 | ±0·62 | $N\pi\pi$ | | 657 |
| | | | $\sigma = 10{\cdot}4$ | | | | $\Lambda K$ | <16 | 437 |
| | | | | | | | $N\eta$ | ~4 | 545 |
| | $N(1990)$ | 1/2(7/2⁺) $F_{17}$ | $T = 1{\cdot}49$ | 1980 to | 220 to | 3·96 | $N\pi$ | 11 | 766 |
| | | | $p = 1{\cdot}63$ | 2000 | 250 | ±0·47 | $N\pi\pi$ | | 743 |
| | | | $\sigma = 8{\cdot}34$ | | | | | | |
| | $N'''(2040)$ | 1/2(3/2⁻) $D'''_{13}$ | $T = 1{\cdot}60$ | 2030 to | 240 to | 4·16 | $N\pi$ | 17 | 797 |
| | | | $p = 1{\cdot}73$ | 2060 | 290 | ±0·56 | $N\pi\pi$ | | 775 |
| | | | $\sigma = 7{\cdot}70$ | | | | | | |
| | $N(2190)$ | 1/2(7/2⁻) $G_{17}$ | $T = 1{\cdot}94$ | 2000 to | 300 | 4·80 | $N\pi$ | 35 | 888 |
| | | | $p = 2{\cdot}07$ | 2260 | | ±0·65 | $N\pi\pi$ | | 868 |
| | | | $\sigma = 6{\cdot}21$ | | | | | | |
| | $N(2650)$ | 1/2(?⁻) | $T = 3{\cdot}12$ | 2650 | 360 | 7·02 | $N\pi$ | | 1154 |
| | | | $p = 3{\cdot}26$ | | | ±0·95 | $N\pi\pi$ | | 1140 |
| | | | $\sigma = 3{\cdot}67$ | | | | | | |

TABLE 3a (*cont.*)

| Group | Particle or resonance | $I(J^P)$ | $\pi$ or $K$ beam: $T$ (GeV)<br>$p$ (GeV/c)<br>$\sigma = 4\pi\lambda\!\!\!^{-2}$ (mb) | Mass (MeV) | $\Gamma$ (MeV) | $M^2 \pm \Gamma M$ (GeV$^2$) | Decay modes: Partial mode | Fraction % | $p$ or $p_{max}$ (MeV/c) |
|---|---|---|---|---|---|---|---|---|---|
| $N$ | $N(3030)$ | 1/2(?) | $T = 4{\cdot}27$<br>$p = 4{\cdot}41$<br>$\sigma = 2{\cdot}62$ | 3030 | 400 | 9·18<br>±1·21 | $N\pi$<br>$N\pi\pi$ | | 1366<br>1354 |
| $\Delta$ | $\Delta(1236)$ | 3/2(3/2$^+$) $P'_{33}$ | $T = 0{\cdot}195(++)$<br>$p = 0{\cdot}304$<br>$\sigma = 91{\cdot}8$ | 1236·0<br>±0·6<br>$m_0 - m_{++} = 0{\cdot}45 \pm 0{\cdot}85$<br>$m_- - m_{++} = 7{\cdot}9 \pm 6{\cdot}8$ | 120<br>±2 | 1·53<br>±0·15 | $N\pi$<br>$N\pi^+\pi^-$<br>$N\gamma$ | 99·4<br>0<br>~0·6 | 231<br>89<br>262 |
| $\Delta$ | $\Delta(1650)$ | 3/2(1/2$^-$) $S_{31}$ | $T = 0{\cdot}83$<br>$p = 0{\cdot}96$<br>$\sigma = 16{\cdot}4$ | 1620 to<br>1695 | 130 to<br>250 | 2·72<br>±0·25 | $N\pi$<br>$N\pi\pi$ | 27<br>73 | 547<br>511 |
| $\Delta$ | $\Delta(1670)$ | 3/2(3/2$^-$) $D_{33}$ | $T = 0{\cdot}87$<br>$p = 1{\cdot}00$<br>$\sigma = 15{\cdot}6$ | 1650 to<br>1690 | 175 to<br>300 | 2·79<br>±0·40 | $N\pi$<br>$N\pi\pi$ | 13 | 560<br>525 |
| $\Delta$ | $\Delta(1890)$ | 3/2(5/2$^+$) $F_{35}$ | $T = 1{\cdot}28$<br>$p = 1{\cdot}42$<br>$\sigma = 9{\cdot}88$ | 1840 to<br>1910 | 135 to<br>380 | 3·57<br>±0·52 | $N\pi$<br>$N\pi\pi$ | 17 | 704<br>677 |
| $\Delta$ | $\Delta(1910)$ | 3/2(1/2$^+$) $P_{31}$ | $T = 1{\cdot}33$<br>$p = 1{\cdot}46$<br>$\sigma = 9{\cdot}54$ | 1835 to<br>1935 | 230 to<br>420 | 3·65<br>±0·62 | $N\pi$<br>$N\pi\pi$ | 25 | 716<br>691 |

TABLE 3a (*cont.*)

| | | | | | | | | | |
|---|---|---|---|---|---|---|---|---|---|
| Δ | Δ(1950) | 3/2(7/2⁺) $F_{37}$ | $T = 1{\cdot}41$<br>$p = 1{\cdot}54$<br>$\sigma = 8{\cdot}90$ | 1935 to 1980 | 140 to 220 | 3·80<br>±0·39 | $N\pi$<br>$\Delta(1236)\,\pi$<br>$\Sigma K$<br>$\Sigma(1385)\,K$<br>$\Delta(1236)\,\varrho$ | 45<br>~50<br>2·4<br>1·4<br>seen | 741<br>571<br>460<br>232 |
| | Δ(2420) | 3/2(11/2⁺) | $T = 2{\cdot}50$<br>$p = 2{\cdot}64$<br>$\sigma = 4{\cdot}68$ | 2420 | 310 | 5·86<br>±0·75 | $N\pi$<br>$N\pi\pi$ | 11<br>>20 | 1023<br>1006 |
| | Δ(2850) | 3/2(?⁺) | $T = 3{\cdot}71$<br>$p = 3{\cdot}85$<br>$\sigma = 3{\cdot}05$ | 2850 | 400 | 8·12<br>±1·14 | $N\pi$<br>$N\pi\pi$ | | 1266<br>1254 |
| | Δ(3230) | 3/2(?) | $T = 4{\cdot}94$<br>$p = 5{\cdot}08$<br>$\sigma = 2{\cdot}25$ | 3230 | 440 | 10·4<br>±1·4 | $N\pi$<br>$N\pi\pi$ | | 1475<br>1464 |

TABLE 3b. Baryons $S \neq 0$

| | Particle or resonance | $I(J^P)$ | $\pi$ or K beam: $T$ (GeV<br>$p$ (GeV/c)<br>$\sigma = 4\pi\lambda\!\!\!^{-2}$ (mb) | Mass (MeV) | $\Gamma$ (MeV) | $M^2 \pm \Gamma M$ (GeV²) | Decay modes: Partial mode | Decay modes: Fraction % | $p$ or $p_{max}$ (MeV/c) |
|---|---|---|---|---|---|---|---|---|---|
| | $\Lambda$ | $0(1/2^+)$ | | 1115·6 | | 1·24 | See Table 1 | | |
| | $\Lambda$ (1405) | $0(1/2^-)$ $S_{01}$ | $p < 0$ $K^-p$ | 1405<br>±5 | 40<br>±10 | 1·97<br>±0·06 | $\Sigma\pi$ | 100 | 142 |
| | $\Lambda'$(1520) | $0(3/2^-)$ $D'_{03}$ | $p = 0·389$<br>$\sigma = 84·5$ | 1518<br>±2 | 16<br>±2 | 2·30<br>±0·02 | $N\bar{K}$<br>$\Sigma\pi$<br>$\Lambda\pi\pi$<br>$\Lambda\gamma$<br>$\Sigma\pi\pi$ | 46 ± 1<br>41 ± 1<br>9·6 ±.6<br>0·8 ±.2<br>1·0 ±.2 | 237<br>260<br>252<br>351<br>144 |
| $\Lambda$ | $\Lambda'$(1670) | $0(1/2^-)$ $S'_{01}$ | $p = 0·74$<br>$\sigma = 28·5$ | 1670 | 30 | 2·79<br>±0·05 | $N\bar{K}$<br>$\Lambda\eta$<br>$\Sigma\pi$ | 15<br>35<br>50 | 410<br>66<br>393 |
| | $\Lambda''$(1690) | $0(3/2^-)$ $D''_{03}$ | $p = 0·78$<br>$\sigma = 26·1$ | 1690 | 27 to<br>85 | 2·86<br>±0·07 | $N\bar{K}$<br>$\Sigma\pi$<br>$\Lambda\pi\pi$<br>$\Sigma\pi\pi$ | 20<br>55<br>~15<br>~10 | 429<br>409<br>415<br>352 |
| | $\Lambda$ (1815) | $0(5/2^+)$ $F_{05}$ | $p = 1·05$<br>$\sigma = 16·7$ | 1815<br>±5 | 75<br>±10 | 3·30<br>±0·13 | $N\bar{K}$<br>$\Sigma\pi$<br>$\Sigma$(1385)$\pi$ | 65 ± 1<br>11 ± 1<br>17 ± 3 | 537<br>504<br>358 |

TABLE 3b (*cont.*)

| | | | | | | | | | |
|---|---|---|---|---|---|---|---|---|---|
| Λ | Λ (1830) | 0(5/2⁻) $D_{05}$ | $p = 1{\cdot}09$ | 1835 | 66 to | 3·37 | $N\bar{K}$ | 10 | 550 |
| | | | $\sigma = 15{\cdot}8$ | | 145 | ±0·18 | $\Sigma\pi$ | 30 | 515 |
| | Λ (2100) | 0(7/2⁻) $G_{07}$ | $p = 1{\cdot}68$ | 2100 | 40 to | 4·41 | $N\bar{K}$ | 25 | 748 |
| | | | $\sigma = 8{\cdot}68$ | | 145 | ±0·21 | $\Sigma\pi$ | ~ 1 | 699 |
| | | | | | | | $\Lambda\eta$ | < 3 | 617 |
| | | | | | | | $\Xi K$ | ~ ·5 | 483 |
| | | | | | | | $\Lambda\omega$ | <10 | 443 |
| | Λ (2350) | 0(?) | $p = 2{\cdot}29$ | 2350 | 150 | 5·52 | $N\bar{K}$ | | 913 |
| | | | $\sigma = 5{\cdot}85$ | | | ±0·35 | | | |
| Σ | Σ | 1(1/2⁺) | | (+)1189·4 | | 1·41 | see Table 1 | | |
| | | | | (0)1192·5 | | 1·42 | | | |
| | | | | (−)1197·3 | | 1·43 | | | |
| | Σ (1385) | 1(3/2⁺) $P_{13}$ | $p < 0\ K^-p$ | (+)1383 ± 1 | (+)36 ± 3 | 1·92 | $\Lambda\pi$ | 90 ± 3 | 208 |
| | | | | | | ±0·05 | $\Sigma\pi$ | 10 ± 3 | 117 |
| | | | | (−)1386 ± 2 | (−)36 ± 6 | | | $S = 1{\cdot}4$* | |
| | Σ (1670) | 1(3/2⁻) $D_{13}$ | $p = 0{\cdot}74$ | 1670 | 50 | 2·79 | $N\bar{K}$ | 8 | 410 |
| | | | $\sigma = 28{\cdot}5$ | | | ±0·08 | $\Sigma\pi$ | 50 | 387 |
| | | The branching ratios as reported here are from | | | | | $\Lambda\pi$ | 32 | 447 |
| | | formation experiments. Production experiments | | | | | $\Sigma\pi\pi$ | <14 | 326 |
| | | still confused. | | | | | $\Lambda(1405)\pi$ | < 6 | 207 |
| | | | | | | | $\Lambda\pi\pi$ | <11 | 397 |

TABLE 3b. (*cont.*)

$\Sigma$

| Particle or resonance | $I\,(J^P)$ | $\pi$ or $K$ beam: $T$ (GeV) $p$ (GeV/c) $\sigma = 4\pi\lambda\!\!\!^{-2}$ (mb) | Mass (MeV) | $\Gamma$ (MeV) | $M^2 \pm \Gamma M$ (GeV$^2$) | Decay modes: Partial mode | Fraction % | $p$ or $p_{max}$ (MeV/c) |
|---|---|---|---|---|---|---|---|---|
| $\Sigma(1750)$ | $\underline{1(1/2^-)}\ S_{11}$ | $p = 0{\cdot}91$ | 1750 | 80 | 3·06 | $N\bar{K}$ | ~15 | 483 |
| | | $\sigma = 20{\cdot}7$ | | | ±0·14 | $\Lambda\pi$ | ~20 | 507 |
| | | | | | | $\Sigma\eta$ | seen | 55 |
| $\Sigma(1765)$ | $\underline{1(5/2^-)}\ D_{15}$ | $p = 0{\cdot}94$ | 1765 | 60 to | 3·12 | $N\bar{K}$ | $45 \pm 1$ | 496 |
| | | $\sigma = 19{\cdot}6$ | ±5 | 146 | ±0·21 | $\Lambda\pi$ | $15 \pm 2$ | 518 |
| | | | | | | $\Lambda(1520)\pi$ | $15 \pm 2$ | 187 |
| | | | | | | $\Sigma(1385)\pi$ | $13 \pm 2$ | 315 |
| | | | | | | $\Sigma\pi$ | $\sim 1$ | 461 |
| $\Sigma(1915)$ | $\underline{1}(5/2^+)\ F_{15}$ | $p = 1{\cdot}25$ | 1910 | 50 | 3·65 | $N\bar{K}$ | 10 | 616 |
| | | $\sigma = 13{\cdot}0$ | | | ±0·10 | $\Lambda\pi$ | 5 | 622 |
| Formation and production experiments do not agree. | | | | | | $\Sigma\pi$ | 0·4 | 571 |
| $\Sigma(2030)$ | $\underline{1(7/2^+)}\ F_{17}$ | $p = 1{\cdot}52$ | 2030 | 80 to | 4·12 | $N\bar{K}$ | 10 | 700 |
| | | $\sigma = 9{\cdot}93$ | | 170 | ±0·24 | $\Lambda\pi$ | 35 | 700 |
| | | | | | | $\Sigma\pi$ | 5 | 652 |
| | | | | | | $\Xi K$ | $< 2$ | 412 |
| $\Sigma(2250)$ | $\underline{1}(?)$ | $p = 2{\cdot}04$ | 2250 | 200 | 5·06 | $N\bar{K}$ | | 849 |
| | | $\sigma = 6{\cdot}76$ | | | ±0·45 | | | |
| $\Sigma(2455)$ | $\underline{1}(?)$ | $p = 2{\cdot}57$ | 2455 | 100 | 6·03 | $N\bar{K}$ | | 979 |
| | | $\sigma = 5{\cdot}09$ | | | ±0·25 | | | |

TABLE 3b. (*cont.*)

| | | | | | | | | | |
|---|---|---|---|---|---|---|---|---|---|
| $\Sigma$ | $\Sigma$ (2595) | 1(?) | $p = 2{\cdot}95$ <br> $\sigma = 4{\cdot}30$ | 2595 | ~140 | 6·73 <br> ±0·36 | $N\bar{K}$ | | 1064 |
| $\Xi$ | $\Xi$ | 1/2(1/2⁺) | | (0)1314·7 <br> (−)1321·3 | | 1·73 <br> 1·75 | see Table 1 | | |
| $\Xi$ | $\Xi$ (1530) | 1/2(3/2⁺) <br> *p*-wave | | (0)1528·9 ± 1·1 <br> (−)1533·8 ± 1·7 | 7·3 <br> ±1·7 | 2·34 <br> ±0·01 | $\Xi\pi$ | 100 | 144 |
| $\Xi$ | $\Xi$ (1820) | 1/2(?) | | 1820 | ~30 | 3·31 | $\Lambda\bar{K}$ | 30 | 396 |
| | | | | | | ±0·05 | $\Xi\pi$ | 10 | 413 |
| | | | | | | | $\Xi$ (1530)$\pi$ | 30 | 234 |
| | | | | | | | $\Sigma\bar{K}$ | 30 | 306 |
| $\Xi$ | $\Xi$ (1930) | 1/2(?) | | 1930 | 110 | 3·72 | $\Xi\pi$ | large | 499 |
| | | | | | | ±0·21 | $\Lambda\bar{K}$ | small | 502 |
| $\Xi$ | $\Xi$ (2030) | 1/2(?) | | 2030 | 50 | 4·12 | $\Xi\pi$ | small | 573 |
| | | | | | | ±0·11 | $\Lambda\bar{K}$ | ~20 | 587 |
| | | | | | | | $\Sigma\bar{K}$ | ~70 | 524 |
| | | | | | | | $\Xi$ (1530)$\pi$ | small | 421 |
| $\Xi$ | $\Xi$ (2250) | 1/2(?) | | 2250 | 130 | 5·06 | $\Lambda\bar{K}\pi$ | seen | 689 |
| | | | | | | 0·29 | $\Sigma\bar{K}\pi$ | seen | 631 |
| | | | | | | | $\Xi\pi\pi$ | seen | 701 |
| $\Xi$ | $\Xi$ (2500) | 1/2(?) | | 2500 | 60 | 6·25 | $\Xi\pi\pi$ | seen | 839 |
| | | | | | | 0·15 | $\Lambda\bar{K}\pi$ | seen | 839 |
| | $\Omega^-$ | 0(3/2⁺) | | 1672·4 | | 2·80 | see Table 1 | | |

## TABLE 4

### GENERAL ATOMIC AND NUCLEAR CONSTANTS‡

| | |
|---|---|
| $N$ | $= 6{\cdot}022169(40) \times 10^{23}$ mole$^{-1}$ |
| $c$ | $= 2{\cdot}997925(10) \times 10^{10}$ cm sec$^{-1}$ |
| $e$ | $= 4{\cdot}803250(21) \times 10^{-10}$ esu $= 1{\cdot}6021917(70) \times 10^{-19}$ coulomb |
| 1 MeV | $= 1{\cdot}6021917(70) \times 10^{-6}$ erg |
| $\hbar$ | $= 6{\cdot}582183(22) \times 10^{-22}$ MeV sec |
| | $= 1{\cdot}0545919(80) \times 10^{-27}$ erg sec |
| $\hbar c$ | $= 1{\cdot}9732891(66) \times 10^{-11}$ MeV cm $= 197{\cdot}32891(66)$ MeV fermi |
| $\alpha$ | $= e^2/\hbar c = 1/137{\cdot}03602(21)$ |
| $k_{\text{Boltzmann}}$ | $= 1{\cdot}380622(59) \times 10^{-16}$ erg K$^{-1}$ |
| | $= 8{\cdot}61708(37) \times 10^{-11}$ MeV K$^{-1}$ $= 1$ eV/11604·85(49) K |
| $m_e$ | $= 0{\cdot}5110041(16)$ MeV $= 9{\cdot}109558(54) \times 10^{-31}$ kg |
| $m_p$ | $= 938{\cdot}2592(52)$ MeV $= 1836{\cdot}109(11)\, m_e = 6{\cdot}72211(63)\, m_{\pi^\pm}$ |
| | $= 1{\cdot}00727661(8)\, m_1$ (where $m_1 = 1$ amu $= 1/12\, m_{C^{12}} = 931{\cdot}4812(52)$ MeV) |
| $r_e$ | $= e^2/m_e c^2 = 2{\cdot}817939(13)$ fermi ((1 fermi $= 10^{-13}$ cm) |
| $\lambda\!\!\!^{-}_e$ | $= \hbar/m_e c = r_e \alpha^{-1} = 3{\cdot}861592(12) \times 10^{-11}$ cm |
| $a_{\infty\text{Bohr}}$ | $= \hbar^2/m_e e^2 = r_e \alpha^{-2} = 0{\cdot}52917715(81)$ A (1 A $= 10^{-8}$ cm) |
| $\sigma_{\text{Thomson}}$ | $= 8/3\, \pi r_e^2 = 0{\cdot}6652453(61) \times 10^{-24}$ cm$^2$ $= 0{\cdot}6652453(61)$ barns |
| $\mu_{\text{Bohr}}$ | $= e\hbar/2m_e c = 0{\cdot}5788381(18) \times 10^{-14}$ MeV gauss$^{-1}$ |
| $\mu_{\text{nucleon}}$ | $= e\hbar/2m_p c = 3{\cdot}152526(21) \times 10^{-18}$ MeV gauss$^{-1}$ |
| $\frac{1}{2}\omega^e{}_{\text{cyclotron}}$ | $= e/2m_e c = 8{\cdot}794014(27) \times 10^6$ rad sec$^{-1}$ gauss$^{-1}$ |
| $\frac{1}{2}\omega^p{}_{\text{cyclotron}}$ | $= e/2m_p c = 4.789484(27) \times 10^3$ rad sec$^{-1}$ gauss$^{-1}$ |

Hydrogen-like atom (non-relativistic, $\mu$ = reduced mass):

$$\left(\frac{v}{c}\right)_{\text{rms}} = \frac{Ze^2}{n\hbar c}\ ;\ E_n = \frac{\mu}{2} v^2 = \frac{\mu Z^2 e^4}{2(n\hbar)^2}\ ;\ a_n = \frac{n^2\hbar^2}{\mu Z e^2}$$

$R_\infty = m_e e^4/2\hbar^2 = m_e c^2 \alpha^2/2 = 13{\cdot}605826(45)$ eV (Rydberg)

$pc = 0{\cdot}3\, H\varrho$(MeV, kilogauss, cm); 0·3 (which is $10^{-11}\, c$) enters because there are $\approx$ 300 "volts"/esu volt.

| | |
|---|---|
| 1 year (sidereal) | $= 365{\cdot}256$ days $= 3{\cdot}1557 \times 10^7$ sec ($\sim \pi \times 10^7$ sec) |
| density of dry air | $= 1{\cdot}205$ mg cm$^{-3}$ (at 20 °C, 760 mm) |
| acceleration by gravity | $= 980{\cdot}62$ cm sec$^{-2}$ (sea level, 45°) |
| gravitational constant | $= 6{\cdot}6732(31) \times 10^{-8}$ cm$^3$g$^{-1}$sec$^{-2}$ |
| 1 calorie (thermochemical) | $= 4{\cdot}184$ joules |
| 1 atmosphere | $= 1033{\cdot}2275$ g cm$^{-2}$ |
| 1 eV per particle | $= 11604{\cdot}85(49)$ °K (from $E = kT$) |

NUMERICAL CONSTANTS

| | | | |
|---|---|---|---|
| $\pi$ | $= 3{\cdot}1415927$ | 1 rad | $= 57{\cdot}2957795$ deg |
| $e$ | $= 2{\cdot}7182818$ | $1/e$ | $= 0{\cdot}3678794$ |
| ln 2 | $= 0{\cdot}6931472$ | ln 10 | $= 2{\cdot}3025851$ |
| $\log_{10} 2$ | $= 0{\cdot}3010300$ | $\log_{10} e$ | $= 0{\cdot}4342945$ |

‡ Compiled by Stanley J. Brodsky, based mainly on the adjustment of the fundamental physical constants by B. N. Taylor, W. H. Parker, and D. N. Langenberg, *Rev. Mod. Phys.* **41**, 375 (1969). The figures in parentheses correspond to the 1 standard deviation uncertainty in the last digits of the main number.

## A.2. Units

In the equations for quantum mechanical systems, the symbols $\hbar$ (Planck's constant/$2\pi$) and $c$ (the velocity of light) occur repeatedly. It is therefore convenient to adopt a system of units which permits the elimination of these symbols from the quantum mechanical equations.

This result may be achieved by employing a system which reduces all quantities in the c.g.s. system to powers of $\hbar$ and $c$ and one natural dimension – length. Thus consider a quantity $A'$ with the following dimensions in the c.g.s. system:

$$[A'] = M^{\alpha}L^{\beta}T^{\gamma}.$$

In order to introduce the new quantity $A$ with dimensions of length, $L$, we must combine $\hbar$ and $c$ in powers such that the dimensions of $M$ and $T$ are zero. Now in the c.g.s. system $\hbar$ and $c$ possess the following dimensions:

$$[\hbar] = ML^2T^{-1}, \qquad [c] = LT^{-1}.$$

Therefore

$$[A] = [A']\,f(\hbar, c) = [A']\,[\hbar]^{\delta}\,[c]^{\varepsilon}$$
$$= M^{\alpha}L^{\beta}T^{\gamma}(ML^2T^{-1})^{\delta}\,(LT^{-1})^{\varepsilon}$$

and to set $M$ and $T$ to zero powers we must have the following equations

$$[M] \qquad \alpha + \delta = 0 \qquad (\delta = -\alpha) \tag{A.2.1}$$
$$[T] \qquad \gamma - \delta - \varepsilon = 0 \qquad (\varepsilon = \alpha + \gamma)$$

and so

$$[A] = [A']\,\hbar^{-\alpha}c^{\alpha+\gamma}$$
$$= L^{\beta-\alpha+\gamma}. \tag{A.2.2}$$

Thus mass, for example, has dimensions (length)$^{-1}$ in the new system. Similarly, since energy has dimensions,

$$[E'] = ML^2T^{-2} \equiv M^{\alpha}L^{\beta}T^{\gamma}$$

in c.g.s. units, then

$$[E] = L^{2-1-2} = L^{-1}, \qquad E \equiv E'/\hbar c \tag{A.2.3}$$

so that

$$E'_{\mathrm{ergs}} = E_{\mathrm{natural\ units}} \times 1{\cdot}05 \times 10^{-27} \times 3 \times 10^{10}.$$

It is a straightforward matter to show that $\hbar = c = 1$ in natural units.

Now let us examine electronic charge, in electrostatic units

$$[e'] = M^{1/2}L^{3/2}T^{-1}$$

therefore, in natural units

$$[e] = L^0, \qquad e = \frac{e'}{\hbar^{1/2}c^{1/2}} \tag{A.2.4}$$

and

$$e'_{\text{esu}} = e_{\text{natural units}}\,\hbar^{1/2}c^{1/2}.$$

This gives us (approximately)

$$e = \sqrt{\frac{1}{137}}$$

in natural units. However, charge is defined by Coulomb's law through a proportionality factor, and a more convenient unit is provided by a rationalised system in which $e$ has the approximate value

$$e \sim \sqrt{\left(\frac{4\pi}{137}\right)} \quad \text{or} \quad \frac{e^2}{4\pi} \sim \frac{1}{137}. \tag{A.2.5}$$

Numerical data on energy, momentum and mass (and sometimes length) are normally quoted in MeV units. The first three quantities are related by the equation

$$E^2 = p^2c^2 + m^2c^4$$

and momentum and mass have MeV/c and MeV/$c^2$ units respectively (the c is frequently suppressed). The MeV unit is related to other frequently encountered quantities in the following manner

$$1\ \text{eV} = \frac{e_{\text{esu}}}{c} \times 10^8\ \text{erg}$$

$$1\ \text{MeV} = 1{\cdot}6021 \times 10^{-6}\ \text{erg}$$

$$\hbar = 1{\cdot}054 \times 10^{-27}\ \text{erg sec} = 6{\cdot}582 \times 10^{-22}\ \text{MeV sec}$$

$$\hbar c = 1{\cdot}973 \times 10^{-11}\ \text{MeV cm}$$

$$1\ \text{fermi} = 10^{-13}\ \text{cm} = (197\ \text{MeV/c})^{-1}$$

where the last relation arises from the de Broglie equation $\lambda = \hbar/p$.

## A.3. Some Points from Quantum Mechanics

This appendix does not set out to be a summary of a textbook on quantum mechanics. There are many excellent introductions to this subject. Instead we attempt to cover below some of the points on notation used in the text.

### A.3(a). Notation

Wherever possible the Dirac notation has been used. This is a convenient shorthand which displays in a clear fashion the relevant parameters in any physical situation and avoids the labour of writing in integrals. It is based on the use of bra ($|\rangle$) and ket ($\langle|$) vectors and indeed uses many points which are familiar from vector algebra. In this notation an integral like $\int d\mathbf{x}\psi_1^*\psi_2$ is written as $\langle\psi_1 \mid \psi_2\rangle$ and referred to as the scalar or inner product, whilst $|\psi_2\rangle\langle\psi_1|$ is called the outer product.

### A.3(b). States

In quantum mechanics the state or wave function $\psi(x)$ is assumed to contain the maximum information available on a system. It and its complex conjugate have been written as

$$\psi(x) = \langle x \mid \psi\rangle \qquad \psi^*(x) = \langle\psi \mid x\rangle \tag{A.3.1}$$

and the normalisation condition is

$$\int dx\, \psi^*\psi = \int dx\, \langle\psi \mid x\rangle\, \langle x \mid \psi\rangle = \langle\psi \mid \psi\rangle = 1. \tag{A.3.2}$$

The state $|\psi\rangle$ can often be conveniently expanded as a linear set of states (each of which represents a possible state of a particle). Firstly, in terms of wave functions

$$\psi(x) = c_1\psi_1(x) + c_2\psi_2(x) + \cdots = \sum_{i=1}^{n} c_i\psi_i$$

where the coefficients $c$ are complex numbers. In the Dirac notation

$$\langle x \mid \psi\rangle = \sum_{i=1}^{n} c_i\,\langle x \mid \psi_i\rangle.$$

For the time being we shall drop the $x$ variable for convenience of writing. If the expansion can be made in terms of a set of orthogonal basis states

$$|\psi\rangle = \sum_i c_i \, |\alpha_i\rangle \tag{A.3.3}$$

where the orthonormality condition is

$$\langle \alpha_i \mid \alpha_j \rangle = \delta_{ij} \tag{A.3.4}$$

then the normalisation condition (A.3.2), and (A.3.1), gives

$$\langle \psi \mid \psi \rangle = \sum_{i,j} c_i^* c_j \, \langle \alpha_i \mid \alpha_j \rangle = \sum_{i=1}^{n} |c_i|^2 = 1 \tag{A.3.5}$$

and so $|c_i|^2$ is the probability of forming the state $|\alpha_i\rangle$.

For two states $|\phi\rangle$ and $|\psi\rangle$

$$|\phi\rangle = \sum_i d_i \, |\alpha_i\rangle \qquad |\psi\rangle = \sum_i c_i \, |\alpha_i\rangle$$

it is apparent that

$$\langle \psi \mid \phi \rangle = \sum_i c_i^* d_i \qquad \langle \phi \mid \psi \rangle = \sum_i d_i^* c_i .$$

Thus $\langle \psi \mid \phi \rangle$ can be regarded as a scalar product (i.e. a pure number). It is also often called the overlap function; the above equations show that

$$\langle \psi \mid \phi \rangle = \langle \phi \mid \psi \rangle^* . \tag{A.3.6}$$

A convenient way of representing kets and bras are as column and row vectors respectively

$$|\psi\rangle = \begin{matrix} c_1 \\ c_2 \\ \vdots \\ c_n \end{matrix} \qquad \langle \phi | = d_1^* d_2^* \ldots d_n^* \tag{A.3.7}$$

from which equation (A.3.6) immediately follows. It is apparent from this notation that the bra is the Hermitian conjugate (transposed and complex conjugated) of the ket.

An important relation in quantum mechanics is the completeness condition. Equations (A.3.3) to (A.3.6) imply that

$$d_i^* = \langle \phi \mid \alpha_i \rangle \qquad c_j = \langle \alpha_j \mid \psi \rangle \tag{A.3.8}$$

hence

$$\begin{aligned}\langle\phi \mid \psi\rangle &= \sum_{i,j} \langle\phi \mid \alpha_i\rangle \langle\alpha_i \mid \alpha_j\rangle \langle\alpha_j \mid \psi\rangle \\ &= \sum_{i,j} \langle\phi \mid \alpha_i\rangle \langle\alpha_j \mid \psi\rangle \,\delta_{ij} \\ &= \sum_{i} \langle\phi \mid \alpha_i\rangle \langle\alpha_i \mid \psi\rangle\end{aligned}$$

and so the completeness condition is

$$\sum_i |\alpha_i\rangle \langle\alpha_i| = \hat{1} \tag{A.3.9}$$

where $\hat{1}$ is the unit operator.

### A.3(c). Operators

An operator $A$ acts on a ket vector from the left, or a bra from the right and transforms them into another ket or bra

$$A\,|\psi\rangle \rightarrow |\psi'\rangle \qquad \langle\psi|\,A \rightarrow \langle\psi'|\,. \tag{A.3.10}$$

If the ket or bra remains unchanged under the operation $A$ is called the unit operator

$$A\,|\psi\rangle = |\psi\rangle \qquad A = \hat{1}\,. \tag{A.3.11}$$

If the operation is of the form

$$A\,|\psi\rangle = a\,|\psi\rangle \tag{A.3.12}$$

then $|\psi\rangle$ is said to be an eigenstate of $A$ with eigenvalue $a$.

If the states $|\psi\rangle$ and $|\psi'\rangle$ can be expanded as in (A.3.3)

$$|\psi\rangle = \sum_{i=1}^{n} c_i\,|\alpha_i\rangle \qquad |\psi'\rangle = \sum_{i=1}^{n} c_j'\,|\alpha_j\rangle$$

then

$$\begin{aligned}A\,|\psi\rangle &= \sum_i c_i A\,|\alpha_i\rangle = \sum_{i,j} c_i f_{ij}\,|\alpha_j\rangle \\ &= |\psi'\rangle = \sum_j c_j'\,|\alpha_j\rangle\end{aligned}$$

and so

$$c_j' = \sum_i c_i f_{ij}.$$

If the $|\alpha_i\rangle$ are eigenstates of $A$ so that $f_{ij} = a_i\delta_{ij}$, then

$$\langle\psi|\,A\,|\psi\rangle = \sum_i |c_i|^2\,a_i \tag{A.3.13}$$

and it follows from (A.3.5) that the probability that $|\psi\rangle$ contains an eigenvalue $a_i$ is $|c_i|^2$.

The term

$$\langle\alpha_i| A |\alpha_j\rangle = A_{ij} \tag{A.3.14}$$

if often called the matrix element of the operator $A$. The conjugate of $A$ is given by $A^\dagger$, where the symbol $\dagger$ implies that matrix elements are transposed and complex conjugated

$$\langle\alpha_i| A^\dagger |\alpha_j\rangle = (\langle\alpha_j| A |\alpha_i\rangle)^*. \tag{A.3.15}$$

If an operator is Hermitian, $A = A^\dagger$, then its eigenvalues must be real, for if $A\,|\,\alpha_i\rangle = a\,|\,\alpha_i\rangle$ then

$$\begin{aligned}\langle\alpha_i| A |\alpha_i\rangle &= a_i\\ = \langle\alpha_i| A^\dagger |\alpha_i\rangle &= (\langle\alpha_i| A |\alpha_i\rangle)^* = a_i^*\end{aligned}$$

and so $a_i = a_i^*$. Since observable quantities are real numbers it follows that the associated operators must be Hermitian.

## A.3(d). Common situations

Since the Dirac notation is simply a convenient shorthand‡ for well-explored quantum mechanical situations, we give here a few examples of commonly occurring terms in quantum mechanics and their forms in the Dirac notation ($Y$ and $j$ are a spherical harmonic and Bessel function respectively)

$$\begin{aligned}\langle x| \psi\rangle &= \psi(x) & \langle lmr\,|\,\psi\rangle &= \psi_{lm}(r)\\ \langle lm| \vartheta\phi\rangle &= Y_l^m(\vartheta,\phi) & \langle l|\,kr\rangle &= j_l(k,r).\end{aligned} \tag{A.3.16}$$

The completeness relation (A.3.9) then allows us to make expansions of the type

$$|\vartheta\phi\rangle = \sum_{l,m} |lm\rangle\,\langle lm\,|\,\vartheta\phi\rangle = \sum_{l,m} Y_l^m(\vartheta,\phi)\,|lm\rangle.$$

‡ In addition to being a shorthand it allows us to express the relevant physical variables in a very clear way.

**A.3(e). Clebsch–Gordan coefficients**

A further completeness relation which is often used in particle physics occurs in the expansion of angular momentum (or isospin) states for particles, say 1 and 2 into states of total angular momentum $|JM\rangle$ with Clebsch–Gordan coefficients $C$

$$\begin{aligned} |j_1m_1; j_2m_2\rangle &= \sum_{J,M} |JM\rangle \langle JM| j_1m_1; j_2m_2\rangle \\ &= \sum_J C_{J\ j_1\ j_2}^{M\ m_1\ m_2} |JM\rangle \end{aligned} \tag{A.3.17}$$

where the summation over $M$ has been dropped since $M = m_1 + m_2$ from the conservation laws. These also limit $J$ to values between $|j_1 - j_2|$ and $(j_1 + j_2)$. The reverse expansion can also be made

$$|JM\rangle = \sum_{m_1=-j}^{m_1=j_2} \sum_{m_2=M-m_1} C_{J\ j_1\ j_2}^{M\ m_1\ m_2} |j_1m_1; j_2m_2\rangle. \tag{A.3.18}$$

The $C$s are the Clebsch–Gordan coefficients. They can be evaluated by repeated application of raising and lowering operators (*P.E.P.*, p. 197); numerical values are given in the accompanying tables for expansions involving $j_2 = \frac{1}{2}$, 1. A typical situation in which they are used is the expansion of the $\pi^-p$ system into its isospin states. The $\pi^-$-meson has $i = 1$, $i_3 = -1$ and the proton $i = \frac{1}{2}$, $i_3 = \frac{1}{2}$, thus $I = \frac{3}{2}, \frac{1}{2}$ and $I_3 = -\frac{1}{2}$ for the combined system. Consultation of Table A.3.1 yields

$$\begin{aligned} |\pi^-p\rangle &= \sum_{I=1/2}^{3/2} C_{I\ \ 1\ \ 1/2}^{-1/2\ -1\ 1/2} |I, \tfrac{1}{2}\rangle \\ &= \sqrt{\tfrac{1}{3}}\, |\tfrac{3}{2}\rangle - \sqrt{\tfrac{2}{3}}\, |\tfrac{1}{2}\rangle. \end{aligned}$$

TABLE A.3.1

| | $C_{J\ j_1\ \frac{1}{2}}^{M\ m_1\ m_2}$ | |
|---|---|---|
| | $m_2 = \frac{1}{2}$ | $m_2 = -\frac{1}{2}$ |
| $J = j_1 + \frac{1}{2}$ | $\left[\dfrac{j_1 + M + \frac{1}{2}}{2j_1 + 1}\right]^{\frac{1}{2}}$ | $\left[\dfrac{j_1 - M + \frac{1}{2}}{2j_1 + 1}\right]^{\frac{1}{2}}$ |
| $J = j_1 - \frac{1}{2}$ | $-\left[\dfrac{j_1 - M + \frac{1}{2}}{2j_1 + 1}\right]^{\frac{1}{2}}$ | $\left[\dfrac{j_1 + M + \frac{1}{2}}{2j_1 + 1}\right]^{\frac{1}{2}}$ |

TABLE A.3.2

| | $C^{M\,m_1\,m_2}_{J\;j_1\;1}$ | | |
|---|---|---|---|
| | $m_2 = 1$ | $m_2 = 0$ | $m_2 = -1$ |
| $J = j_1 + 1$ | $\left[\frac{(j_1+M)(j_1+M+1)}{(2j_1+1)(2j_1+2)}\right]^{\frac{1}{2}}$ | $\left[\frac{(j_1-M+1)(j_1+M+1)}{(2j_1+1)(j_1+1)}\right]^{\frac{1}{2}}$ | $\left[\frac{(j_1-M)(j_1-M+1)}{(2j_1+1)(2j_1+2)}\right]^{\frac{1}{2}}$ |
| $J = j_1$ | $-\left[\frac{(j_1+M)(j_1-M+1)}{2j_1(j_1+1)}\right]^{\frac{1}{2}}$ | $\frac{M}{[j_1(j_1+1)]^{\frac{1}{2}}}$ | $\left[\frac{(j_1-M)(j_1+M+1)}{2j_1(j_1+1)}\right]^{\frac{1}{2}}$ |
| $J = j_1 - 1$ | $\left[\frac{(j_1-M)(j_1-M+1)}{2j_1(2j_1+1)}\right]^{\frac{1}{2}}$ | $-\left[\frac{(j_1-M)(j_1+M)}{j_1(2j_1+1)}\right]$ | $\left[\frac{(j_1+M+1)(j_1+M)}{2j_1(2j_1+1)}\right]^{\frac{1}{2}}$ |

## A.3(f). Spherical functions

A plane wave of wave number $k$ (where $k = \hbar/p$), travelling in the positive $z$-direction, can be represented by the function $e^{ikz}$. This function can be expanded into a sum of spherical waves in the following manner:

$$e^{ikz} = \sum_{l=0}^{\infty} \sqrt{[4\pi(2l+1)]}\, i^l j_l(kr)\, Y_l^0(\theta, \phi) \tag{A.3.19}$$

and if an arbitrary direction $\theta', \phi'$ is chosen for the direction of the plane wave, then

$$e^{i\mathbf{k}\cdot\mathbf{r}} = 4\pi \sum_{l=0}^{\infty} \sum_{m=-l}^{m=+l} i^l j_l(kr)\, Y_l^{m*}(\theta', \phi')\, Y_l^m(\theta, \phi)$$

(Blatt and Weisskopf, 1952). In these equations $j_l(kr)$ is a spherical Bessel function and $Y_l^0(\theta, \phi)$ is a spherical harmonic. We list below some useful properties of these functions.

## A.3(g). Spherical Bessel functions

The spherical Bessel function may be written as

$$j_l(\mu) = \left(\frac{\pi}{2\mu}\right)^{\frac{1}{2}} J_{l+\frac{1}{2}}(\mu) \tag{A.3.20}$$

where $J$ is an ordinary Bessel function and $l$ is an integer (tables of Bessel functions may be found in Jahnke and Emde, 1945). The spherical function may be expressed as

$$j_l(\mu) = (-\mu)^l \left(\frac{1}{\mu}\frac{d}{d\mu}\right)^l \left(\frac{\sin\mu}{\mu}\right) \tag{A.3.21}$$

and so

$$j_0(\mu) = \frac{\sin\mu}{\mu}, \qquad j_1(\mu) = \frac{\sin\mu}{\mu^2} - \frac{\cos\mu}{\mu}$$

$$j_2(\mu) = \left(\frac{3}{\mu^3} - \frac{1}{\mu}\right)\sin\mu - \frac{3}{\mu^2}\cos\mu.$$

For small and large values of $\mu$ the function behaves like

$$j_l(\mu) = \frac{\mu^l}{(2l+1)!!} \qquad |\mu| \ll l$$

$$j_l(\mu) = \frac{1}{\mu}\sin\left(\mu - l\frac{\pi}{2}\right) \qquad |\mu| \gg l$$

where

$$(2l+1)!! = 1 \times 3 \times 5, \ldots (2l+1).$$

## A.3(h). Spherical harmonics

The spherical harmonics $Y_l^m(\theta, \phi)$ satisfy the relations

$$Y_l^m(\theta, \phi) = \left[\frac{2l+1}{4\pi}\frac{(l-m)!}{(l+m)!}\right]^{\frac{1}{2}} P_l^m(\cos\theta)\, e^{im\phi} \tag{A.3.22}$$

$$\begin{aligned} \mathbf{L}^2 Y_l^m &= l(l+1)\, Y_l^m \\ L_z Y_l^m &= m Y_l^m \\ L_\pm Y_l^m &= (L_x \pm iL_y)\, Y_l^m = [l(l+1) - m(m \pm 1)]^{1/2}\, Y_l^{m\pm 1} \end{aligned} \tag{A.3.23}$$

In the above relations

$$l = 0, 1, 2, 3, \ldots$$

$$m = -l, (-l+1), \ldots (l-1), l$$

and $P_l^m$ is an associated Legendre polynomial. For positive $m$

$$P_l^m(x) = (-1)^m (1 - x^2)^{m/2} \frac{d^m}{dx^m} P_l(x)$$

$$P_l(x) = \frac{1}{2^l l!} \frac{d^l}{dx^l} (x^2 - 1)^l.$$

In the above equations we have followed the phase convention of Condon and Shortley (1935); most works use this convention.

The following relationships exist between the components $\pm m$ for a given value of $l$

$$\begin{aligned} P_l^{-m}(x) &= (-1)^m \frac{(l-m)!}{(l+m)!} P_l^m(x) \\ Y_l^{-m}(\theta, \phi) &= (-1)^m Y_l^{m*}(\theta, \phi). \end{aligned} \quad (A.3.24)$$

Using (A.3.22) the lowest order spherical harmonics are given by

$$l = 0 \quad Y_0^0 = \frac{1}{\sqrt{(4\pi)}}$$

$$l = 1 \quad Y_1^0 = \sqrt{\left(\frac{3}{4\pi}\right)} \cos\theta \qquad Y_1^1 = -\sqrt{\left(\frac{3}{8\pi}\right)} \sin\theta\, e^{i\phi}$$

$$l = 2 \quad Y_2^0 = \sqrt{\left(\frac{5}{16\pi}\right)} (3\cos^2\theta - 1) \quad (A.3.25)$$

$$Y_2^1 = -\sqrt{\left(\frac{15}{8\pi}\right)} \sin\theta \cos\theta\, e^{i\phi}$$

$$Y_2^2 = \sqrt{\left(\frac{15}{32\pi}\right)} \sin^2\theta\, e^{2i\phi}$$

$$l = 3 \quad Y_3^0 = \sqrt{\left(\frac{7}{16\pi}\right)} (5\cos^3\theta - 3\cos\theta)$$

$$Y_3^1 = -\sqrt{\left(\frac{21}{64\pi}\right)} \sin\theta (5\cos^2\theta - 1)\, e^{i\phi}$$

$$Y_3^2 = \sqrt{\left(\frac{105}{32\pi}\right)} \sin^2\theta \cos\theta \, e^{2i\phi}$$

$$Y_3^3 = -\sqrt{\left(\frac{35}{64\pi}\right)} \sin^3\theta \, e^{3i\phi}.$$

Values for $Y_l^{-m}(\theta, \phi)$ can be obtained from (A.3.24).

The function $Y_l^m$ satisfies the following conditions

$$\int_0^{2\pi} d\phi \int_0^{\pi} d\theta \sin\theta \, Y_{l'}^{m'*}(\theta, \phi)\, Y_l^m(\theta, \phi) = \delta_{ll'}\delta_{mm'}$$

$$\sum_{l=0}^{\infty} \sum_{m=-l}^{m=+l} Y_l^{m*}(\theta', \phi')\, Y_l^m(\theta, \phi) = \delta(\phi - \phi')\, \delta(\cos\theta - \cos\theta'). \qquad \text{(A.3.26)}$$

## A.3(i). Projection operators

It is often desirable to extract an expression relating to a specific angular momentum or isospin state from one which represents the effects of a combination of such states. This process can be achieved with the aid of projection operators.

The projection operator yields the value of a projection of an abitrary vector in Hilbert space along a specified direction. It possesses the properties

$$P_\alpha |\alpha\rangle = |\alpha\rangle, \qquad P_\alpha |\beta\rangle = 0, \qquad \alpha \neq \beta \qquad \text{(A.3.27)}$$

and satisfies the operator equations

$$\sum_\alpha P_\alpha = \hat{1}, \qquad P_\alpha^2 = P_\alpha, \qquad P_\alpha P_\beta = 0, \qquad \alpha \neq \beta. \qquad \text{(A.3.28)}$$

As an example of the use of projection operators, let us consider the problem of the scattering of spin 0 and spin $\frac{1}{2}$ particles (Lepore, 1950). The total angular momentum operator is

$$\mathbf{J} = (\mathbf{L} + \tfrac{1}{2}\boldsymbol{\sigma})$$

and so

$$\mathbf{J}^2 = \mathbf{L}^2 + \boldsymbol{\sigma}\cdot\mathbf{L} + \tfrac{1}{4}\boldsymbol{\sigma}^2$$

$$\boldsymbol{\sigma}\cdot\mathbf{L} = \mathbf{J}^2 - \mathbf{L}^2 - \tfrac{1}{4}\boldsymbol{\sigma}^2$$

$$\boldsymbol{\sigma}\cdot\mathbf{L}\,|jls\rangle = [j(j+1) - l(l+1) - \tfrac{3}{4}]\,|jls\rangle.$$

Thus if we have two states corresponding to $j = l + \frac{1}{2}$ and $j = l - \frac{1}{2}$, which we shall designate as $|l\,+\rangle$ and $|l\,-\rangle$ respectively, then

$$\begin{aligned} \boldsymbol{\sigma}\cdot\mathbf{L}\,|l\,+\rangle &= [(l+\tfrac{1}{2})(l+\tfrac{3}{2}) - l(l+1) - \tfrac{3}{4}]\,|l\,+\rangle = l\,|l_+\rangle \\ \boldsymbol{\sigma}\cdot\mathbf{L}\,|l\,-\rangle &= -(l+1)\,|l\,-\rangle \end{aligned} \tag{A.3.29}$$

Suitable projection operators for the $|l\,+\rangle$ and $|l\,-\rangle$ states are therefore given by

$$P_+ = \frac{l + 1 + \boldsymbol{\sigma}\cdot\mathbf{L}}{2l+1}, \qquad P_- = \frac{l - \boldsymbol{\sigma}\cdot\mathbf{L}}{2l+1} \tag{A.3.30}$$

since

$$\begin{aligned} P_+\,|l\,+\rangle &= |l\,+\rangle, \qquad & P_-\,|l\,-\rangle &= |l\,-\rangle \\ P_+\,|l\,-\rangle &= 0, & P_-\,|l\,+\rangle &= 0. \end{aligned}$$

Now let us use these operators in a scattering problem. The scattering amplitude for a spinless system was given in (7.1.6) as

$$f(\theta) = \sum_{l=0}^{\infty} (2l+1)\frac{\eta_l - 1}{2ik} P_l^0(\theta) = \sum_l (2l+1)\, f_l P_l^0(\theta).$$

If we introduce spin functions $\chi$ for initial and final states, we may modify the above relation to the following form for scattering between spin states $i$ and $f$:

$$\begin{aligned} f_{fi}(\theta) &= \chi_f^\dagger \sum_l (2l+1)(P_+ f_{l+} + P_- f_{l-})\, P_l^0 \chi_i \\ &= \chi_f^\dagger \sum_l (2l+1)\left(\frac{l+1+\boldsymbol{\sigma}\cdot\mathbf{L}}{2l+1} f_{l+} + \frac{l-\boldsymbol{\sigma}\cdot\mathbf{L}}{2l+1} f_{l-}\right) P_l^0 \chi_i \\ &= \chi_f^\dagger \sum_l \{[(l+1) f_{l+} + l f_{l-}]\, P_l^0 + (f_{l+} - f_{l-})\, \boldsymbol{\sigma}\cdot\mathbf{L} P_l^0\}\, \chi_i. \end{aligned} \tag{A.3.31}$$

Now spin "up" and spin "down" are orthogonal states and $\chi_f^\dagger \chi_i = \delta_{fi}$ for the first expression on the right-hand side of the above equation; thus it represents an amplitude for no spin flip. Now consider the second term; since

$$\begin{aligned} \boldsymbol{\sigma}\cdot\mathbf{L} &= \sigma_1 L_1 + \sigma_2 L_2 + \sigma_3 L_3 \\ &= \tfrac{1}{2}(\sigma_1 - i\sigma_2)(L_1 + iL_2) + \tfrac{1}{2}(\sigma_1 + i\sigma_2)(L_1 - iL_2) + \sigma_3 L_3 \\ &= \sigma_- L_+ + \sigma_+ L_- + \sigma_3 L_3 \end{aligned}$$

where the subscripts plus and minus refer to the raising and lowering operators for angular momentum [see (A.3.23) and *P.E.P.*, p. 193], it is not difficult to see that the second term in (A.3.31) contains the spin flip factor. Consider, for example, the "spin up" state then by (A.3.23)

$$\begin{aligned}\boldsymbol{\sigma}\cdot\mathbf{L}\,P_l^0\begin{pmatrix}1\\0\end{pmatrix} &= (\sigma_- L_+ + \sigma_+ L_- + \sigma_3 L_3)\frac{4\pi}{2l+1}\,Y_l^0\begin{pmatrix}1\\0\end{pmatrix}\\ &= \sigma_- L_+\frac{4\pi}{2l+1}\,Y_l^0\begin{pmatrix}1\\0\end{pmatrix}\\ &= P_l^1\,e^{i\phi}\begin{pmatrix}0\\1\end{pmatrix}.\end{aligned}$$

Thus we see that (A.3.31) contains the nonspin flip ($g$) and spin flip ($h$) terms of § 7.1(d) (since $kf_{l\pm} = T_{l\pm}$)

$$g(\vartheta) = \frac{1}{k}\sum_l\,[(l+1)\,T_{l+} + T_{l-}]\,P_l^0$$

$$h(\vartheta) = \frac{1}{k}\sum_l\,(T_{l+} - T_{l-})\,P_l^1.$$

and when we average over initial spin states and sum over final

$$\frac{d\sigma}{d\Omega} = |g(\vartheta)|^2 + |h(\vartheta)|^2.$$

## A.4. δ-Functions

The Dirac δ-function may be defined by the following properties:

$$\delta(x-a) = 0 \qquad (x \neq a)$$

$$\begin{aligned}\int_b^c dx\,\delta(x-a) &= 1 \quad \text{if } a \text{ is contained in interval } b \text{ to } c \qquad \text{(A.4.1)}\\ &= 0 \quad \text{if } a \text{ is not in this interval.}\end{aligned}$$

The function can be conveniently expressed in mathematical form in the following manner:

$$\frac{1}{2\pi}\int_{-\infty}^{+\infty} dx\, e^{i(k_x-k_x')x} = \delta(k_x - k_x') \qquad \text{(A.4.2)}$$

where $x$ is one-dimensional; the $\delta$-function can also be given multi-dimensional form, for example if $(k - k')x$ is four-dimensional, then

$$\frac{1}{(2\pi)^4}\int_{-\infty}^{+\infty} d^4x\, e^{i(k-k')x} = \delta(k - k') \equiv \delta^4(k - k'). \qquad \text{(A.4.3)}$$

The following useful properties may be associated with the $\delta$-function (Dirac, 1947):

$$\delta(x) = \delta(-x)$$

$$\frac{\partial}{\partial x}\delta(x) = -\frac{\partial}{\partial x}\delta(-x)$$

$$x\,\delta(x) = 0$$

$$x\frac{\partial}{\partial x}\delta(x) = -\delta(x)$$

$$\delta(ax) = \frac{1}{a}\delta(x) \qquad a > 0 \qquad \text{(A.4.4)}$$

$$\delta(x^2 - a^2) = \frac{1}{2a}[\delta(x - a) + \delta(x + a)] \qquad a > 0$$

$$\int dx\, \delta(a - x)\, \delta(x - b) = \delta(a - b)$$

$$f(x)\,\delta(x - a) = f(a)\,\delta(x - a)$$

$$\int dx\, f(x)\, \delta[y(x)] = \sum_i \frac{f(x_i)}{|\partial y(x_i)/\partial x|}$$

where the points $x_i$ are the real roots of $y(x) = 0$ in the interval of integration. The last equation follows from writing

$$z = y(x) \qquad dz = \frac{\partial y}{\partial x}dx$$

hence
$$\int dx\, f(x)\, \delta[y(x)] = \int dz \frac{f(x)}{\partial y/\partial x} \delta(z) = \sum_i \frac{f(x_i)}{|\partial y(x_i)/\partial x|}.$$

Several other singular functions may be associated with the $\delta$-function for example, if a system is discrete rather than continuous the Kroenecke $\delta$-symbol may be used

$$\frac{1}{V} \sum_{\mathbf{x}} e^{i(\mathbf{k}-\mathbf{k}').\mathbf{x}} = \delta_{\mathbf{k}\mathbf{k}'} \tag{A.4.5}$$

$$\delta_{\mathbf{k}\mathbf{k}'} = \begin{cases} 1 & \mathbf{k} = \mathbf{k}' \\ 0 & \mathbf{k} \neq \mathbf{k}'. \end{cases}$$

Other useful functions are the step functions with the properties

$$\theta(t) = \frac{1}{2\pi i} \int_{-\infty}^{+\infty} da \frac{e^{iat}}{a - i\varepsilon} = \begin{cases} 1 & t > 0 \\ 0 & t < 0 \end{cases} \tag{A.4.6}$$

$$\varepsilon(t) = \frac{1}{\pi i} P \int_{-\infty}^{+\infty} \frac{da}{a} e^{iat} = \theta(t) - \theta(-t) = \begin{cases} +1 & t > 0 \\ -1 & t < 0 \end{cases}$$

where $P$ denotes the principal value.

## A.5. Reference Frames and the Lorentz Transformation

The Lorentz transformation relates the components of four vectors in different reference frames (§§ 3.3 and 3.4)

$$p'_\lambda = L_{\lambda\sigma} p_\sigma \tag{A.5.1}$$

where in the case of a transformation along the $x$-axis $L_{\lambda\sigma}$ assumes the form

$$L_{\lambda\sigma} = \begin{pmatrix} \gamma & 0 & 0 & -i\beta\gamma \\ 0 & 1 & 0 & 0 \\ 0 & 0 & 1 & 0 \\ i\beta\gamma & 0 & 0 & \gamma \end{pmatrix} \qquad \gamma = (1 - \beta^2)^{-\frac{1}{2}} \tag{A.5.2}$$

and the components of $p'_\lambda$ are

$$\begin{aligned} p'_x &= \gamma(p_x + \beta E) \\ p'_y &= p_y \\ p'_z &= p_z \\ E' &= \gamma(\beta p_x + E). \end{aligned} \tag{A.5.3}$$

If the transformation is not parallel to an axis the elements $L_{\lambda\sigma}$ become more complicated. They can easily be established by

1. rotating the spatial components of the four vector until it is parallel to the $\boldsymbol{\beta}$ direction; we shall denote the operation by $R(\vartheta, \phi)$;
2. applying the Lorentz transformation (A.5.2);
3. making the reverse rotation $R^{-1}(\vartheta, \phi)$.

Thus

$$L(\boldsymbol{\beta}) = R^{-1}(\vartheta, \phi)\, L(\beta)\, R(\vartheta, \phi) \tag{A.5.4}$$

but in practice working out the matrices is lengthy. Instead the same result can be obtained by considering components of $\mathbf{p}$ parallel ($\|$) and perpendicular ($\perp$) to the direction $\boldsymbol{\beta}$

$$\mathbf{p} = \mathbf{p}_\| + \mathbf{p}_\perp = \boldsymbol{\beta}\frac{\boldsymbol{\beta}\cdot\mathbf{p}}{\beta^2} + \left(\mathbf{p} - \boldsymbol{\beta}\frac{\boldsymbol{\beta}\cdot\mathbf{p}}{\beta^2}\right).$$

Now apply the Lorentz transformation (A.5.2) along $\boldsymbol{\beta}$ and recall that (1) $\boldsymbol{\beta}\cdot\mathbf{p}_\perp = 0$, (2) $\mathbf{p}_\perp$ remains unchanged

$$\begin{aligned} \mathbf{p}' = \mathbf{p}'_\| + \mathbf{p}_\perp &= \gamma(\mathbf{p}_\| + \boldsymbol{\beta}E) + \mathbf{p}_\perp \\ &= \gamma\boldsymbol{\beta}\frac{\boldsymbol{\beta}\cdot\mathbf{p}}{\beta^2} + \gamma\,\boldsymbol{\beta}E + \mathbf{p} - \boldsymbol{\beta}\frac{\boldsymbol{\beta}\cdot\mathbf{p}}{\beta^2} \\ &= \mathbf{p} + \boldsymbol{\beta}\left[(\gamma - 1)\frac{\boldsymbol{\beta}\cdot\mathbf{p}}{\beta^2} + \gamma E\right] \end{aligned}$$

$$E' = \gamma(\boldsymbol{\beta}\cdot\mathbf{p}_\| + E) = \gamma(\boldsymbol{\beta}\cdot\mathbf{p} + E)$$

An inspection of the above equations then shows that the matrix elements of $L$ are

$$\begin{aligned} L_{ij}(\boldsymbol{\beta}) &= \delta_{ij} + \frac{\beta_i \beta_j}{\beta^2}(\gamma - 1) \qquad i, j = 1, 2, 3 \\ L_{i4} &= -L_{4i} = -i\beta_i \gamma \\ L_{44} &= \gamma. \end{aligned} \tag{A.5.5}$$

In practice whenever possible in particle physics quantities are expressed in terms of the scalar products of four-vectors, since these remain invariant under Lorentz transformations.

$$p'_\alpha q'_\alpha = p_\alpha q_\alpha. \tag{A.5.6}$$

Two reference frames are of particular importance in particle physics—the laboratory and centre of momentum or $c$-system. The latter is defined as that system in which the total momentum of the particles is zero. Quantities in the two frames can be readily related by the Lorentz invariance property (A.5.6). Consider a system containing $n$ particles ($n = 1 \to \infty$) with total four momentum $p_{\lambda L}$ in the laboratory frame and $p_{\lambda c}$ in the $c$-system, then in Minkowski notation (§ 2.1)

$$\begin{aligned} p_{\lambda L}^2 &= p_{\lambda c}^2 \qquad \lambda = 1 \to 4 \\ \mathbf{P}_L^2 - E_L^2 &= -E_c^2 \end{aligned} \tag{A.5.7}$$

where we have used the definition $\mathbf{P}_c = 0$. Since $E_c$ has energy but no momentum it is the effective "mass" of the $c$-system and its effective velocity when viewed from the $L$ frame is

$$\boldsymbol{\beta}_c = \frac{\mathbf{P}_L}{E_L}. \tag{A.5.8}$$

If only two particles are present in the $c$-system it is a simple matter to show that their momentum is given by

$$|\mathbf{p}_c| = p_c = \frac{1}{2E_c}\{[E_c^2 - (m_1 + m_2)^2]\,[E_c^2 - (m_1 - m_2)^2]\}^{1/2}. \tag{A.5.9}$$

Another useful Lorentz invariant is the phase space term

$$\frac{d\mathbf{p}}{E} \tag{A.5.10}$$

(this invariance property is obviously not limited to four momenta). The fact that $d\mathbf{p}/E$ is invariant may be readily established by considering a transformation along the $x$-axis and using (A.5.3), then the $y$ and $z$ components remain unchanged and

$$dp'_x = \gamma\left(1 + \beta\frac{\partial E}{\partial p_x}\right)dp_x$$

but

$$E^2 = p_x^2 + p_y^2 + p_z^2 + m^2 \qquad \frac{\partial E}{\partial p_x} = \frac{p_x}{E}$$

and so [with the aid of (A.5.3)]

$$\frac{d\mathbf{p}'}{E'} = \frac{dp'_x\, dp'_y\, dp'_z}{E'} = \frac{\gamma(E + \beta p_x)}{\gamma(\beta p_x + E)}\,\frac{dp_x\, dp_y\, dp_z}{E} = \frac{d\mathbf{p}}{E}.$$

The property of Lorentz invariance of $d\mathbf{p}/E$ is of great importance in the phase space integral

$$N'_q = \frac{1}{(2\pi)^{3q}} \prod_{j=1}^{q} \frac{d\mathbf{p}_j}{E_j}\,\delta(p_f - p_i)$$

since it allows the integral to be cast in the form of a recursion relation (Srivastava and Sudarshan, 1958). Consider a system of $q$ particles with total energy $E$ and zero momentum

$$\begin{aligned} N'_q(0, E) &= \prod_{j=1}^{q} \int \frac{d\mathbf{p}_j}{E_j}\,\delta\left(\sum_j \mathbf{p}_j\right)\delta\left(\sum_j E_j - E\right) \\ &= \int \frac{d\mathbf{p}_q}{E_q} \prod_{j=1}^{q-1} \int \frac{d\mathbf{p}_j}{E_j}\,\delta\left[\sum_{j=1}^{q-1} \mathbf{p}_j - (-\mathbf{p}_q)\right] \\ &\quad \times \delta\left[\sum_{j=1}^{q-1} E_j - (E - E_q)\right] \\ &= \int \frac{d\mathbf{p}_q}{E_j} N'_{q-1}(-\mathbf{p}_q, E - E_q). \end{aligned}$$

Now $N'_{q-1}$ is also Lorentz invariant and so can be evaluated in a system with zero total momentum and energy $\varepsilon$ given by (A.5.7)

$$(E - E_q)^2 - (-\mathbf{p}_q)^2 = \varepsilon^2$$

hence

$$N'_q = \int \frac{d\mathbf{p}_q}{E_q} N'_{q-1}(0, \varepsilon).$$

It is apparent that the reduction can be continued until the two-body state is reached.

One final property of reference frames is worth noting. This is the boost prescription which relates expressions given in a spatially invariant three-dimensional form to Lorentz invariant expressions. The prescription is as follows:

in the rest frame of a system $X$ we have three component vectors $\mathbf{k}$ and energies $\omega$. The boosts induce the following transformations vectors:

$$k_i \to \bar{k}_\mu = k_\mu + X_\mu \frac{(k_\nu X_\nu)}{m_x^2} \tag{A.5.11}$$

pseudo-vectors (subscripts $A$ and $B$ label particles not momenta)

$$(\mathbf{k}_A \times \mathbf{k}_B)_i \to \varepsilon_{\mu\nu\varrho\sigma} k_{A\nu} k_{B\varrho} \frac{X_\sigma}{im_x},$$

energies

$$\omega_A \to -k_{A\mu} \frac{X_\mu}{m_x},$$

Kroenecker delta

$$\delta_{ij} \to \delta_{\mu\nu} + \frac{X_\mu X_\nu}{m_x^2}.$$

Thus in the rest frame of $X$, $X_\nu = 0, im_x$

$$\mu \equiv i = 1, 2, 3 \qquad \bar{k}_i = k_i$$

$$\bar{k}_4 = k_4 + im_x \frac{(k_4\, im_x)}{m_x^2}$$

$$= 0.$$

As an example of the application of the boost prescription consider the matrix element for $\omega$-decay (7.1.29)

$$\begin{aligned} T_{fi} &= i\mathbf{e} \cdot (\mathbf{k}_1 \times \mathbf{k}_2) \\ &\to e_\mu\, \varepsilon_{\mu\nu\varrho\sigma} k_{1\nu} k_{2\varrho} \frac{X_\sigma}{m_x} \\ &= e_\mu\, \varepsilon_{\mu\nu\varrho\sigma} k_{1\nu} k_{2\varrho} (k_{1\sigma} + k_{2\sigma} + k_{3\sigma})/m_x \\ &\equiv \varepsilon_{\mu\nu\varrho\sigma}\, e_\mu\, k_{1\nu} k_{2\varrho} k_{3\sigma} \end{aligned}$$

which is equation (7.1.28). The last line in the above argument follows from the antisymmetry properties of the $\varepsilon$ (3.4.1).

## A.6. γ-Matrices

As we have indicated in Chapter 3 the basic property of the Dirac $\gamma$-matrices is that they satisfy the relation

$$\gamma_\mu \gamma_\lambda + \gamma_\lambda \gamma_\mu = 2\delta_{\lambda\mu}. \tag{A.6.1}$$

It is generally possible to avoid specific representations for the $\gamma$-matrices, although we have given the Dirac–Pauli representation in § 3.5(a). In practice the most frequent evaluation of the matrices is in transition probabilities which involve trace techniques [§ 3.5(h)]; there all that is necessary is equation (A.6.1) together with the fact that the matrices are of $4 \times 4$ dimensions so that

$$\operatorname{tr} \gamma_\lambda^2 = 4 \qquad \lambda = 1 \to 5. \tag{A.6.2}$$

Using these two equations the following properties of the traces can be easily proved ($A_\lambda$, $B_\mu$, $C_\varrho$ and $D_\nu$ are four-vectors; the subscripts run 1 to 5 in conditions (1) to (3)):

(1) $$\operatorname{tr} \gamma_\mu \gamma_\lambda = 4\,\delta_{\mu\lambda}$$

(2) $$\operatorname{tr} \gamma_\lambda \gamma_\mu \gamma_\varrho \gamma_\nu = 4(\delta_{\lambda\mu}\delta_{\varrho\nu} - \delta_{\lambda\varrho}\delta_{\mu\nu} + \delta_{\lambda\nu}\delta_{\mu\varrho})$$

(3) If $(\gamma_\mu \gamma_\lambda \ldots \gamma_\varrho)$ contains an odd number of $\gamma$-matrices (A.6.3)

$$\operatorname{tr} (\gamma_\mu \gamma_\lambda \ldots \gamma_\varrho) = 0$$

(4) $$\operatorname{tr} \gamma_\lambda A_\lambda \gamma_\mu B_\mu = 4\,AB$$

(5) $$\operatorname{tr} \gamma_\lambda A_\lambda \gamma_\mu B_\mu \gamma_\varrho C_\varrho \gamma_\nu D_\nu$$

$$= 4[(AB)(CD) - (AC)(BD) + (AD)(BC)].$$

As an example of an application of the trace technique consider the decay process $\pi^- \to l^-\nu$ discussed in § 6.2(c); here $l^-$ is a charged lepton (muon or electron). We have to evaluate the trace associated with the matrix element (6.2.14)

$$\bar{u}_l(1 + \gamma_5)\, u_\nu.$$

If we make the assumption that the mass of the neutrino is zero, then by equation (3.5.44) the trace, without its normalisation ($2m$) terms, reads

$$\begin{aligned}
&\mathrm{tr}\,[(1+\gamma_5)(-i\gamma p_\nu)\gamma_4(1+\gamma_5)^\dagger\gamma_4(m_l - i\gamma p_l)] \\
&= \mathrm{tr}\,[(1+\gamma_5)(-i\gamma p_\nu)(1-\gamma_5)(m_l - i\gamma p_l)] \\
&= -2i\,\mathrm{tr}\,[(1+\gamma_5)(\gamma p_\nu)(m_l - i\gamma p_l)] \\
&= -8p_\nu p_l \\
&= 8(E_\nu E_l - \mathbf{p}_\nu\cdot\mathbf{p}_l) \\
&= 8[p_l(m_\pi - p_l) + p_l^2] \quad \text{where} \quad p_l = |\mathbf{p}_l|, \quad E_\nu = |\mathbf{p}_\nu| = p_l \\
&= 8m_\pi p_l .
\end{aligned}$$

If we include invariants, normalisation and the phase space factor [§§ 2.5 and 6.2(c)] we then find that the decay rate is

$$\begin{aligned}
\Gamma &\propto |G_A m_l - G_P m_\pi|^2 \frac{m_\pi p_l}{4m_\nu m_l} \frac{m_\nu}{E_\nu} \frac{m_l}{E_l} E_\nu E_l \frac{p_l}{m_\pi} \\
&\propto |G_A m_l - G_P m_\pi|^2\, p_l^2 \\
&\propto |G_A m_l - G_P m_\pi|^2\, (m_\pi^2 - m_l^2)^2
\end{aligned}$$

and the ratio of the decay rates to electron and muon is easily calculated.

## A.7. Symmetry Properties for the Decay $X \to \pi\varrho$

As an example of the determination of the parameters of a resonance, assume that a system decaying to three pions has been found and that two of the pions also form a $\varrho$ resonance.

$$\begin{array}{c} X \\ \pi \leftarrow \cdot \rightarrow \varrho \rightarrow \pi\pi . \end{array} \tag{A.7.1}$$

A final state of 3 pions immediately implies $B = S = 0$. If the resonance is broad ($\Gamma \gg 0$) we can assume that the decay is a strong one and so $G$ parity is conserved (§ 7.2(c)). Thus since there are three pions in the final state

$$G = (-1)^3 = -1. \tag{A.7.2}$$

Since the pion and rho both have isospin 1, the possible values of isospin for $X$ are 0, 1, 2. Systems with isospin 2 have never been observed, so we shall concentrate on 0 and 1. Application of the Clebsch–Gordan coefficients (Table A.3.2) then give the following table.

TABLE A.7.1.

| $I_3$ | $I=1$ | $I=0$ |
|---|---|---|
| 1 | $\frac{1}{\sqrt{2}}(\pi^+\varrho^0 - \pi^0\varrho^+)$ | |
| 0 | $\frac{1}{\sqrt{2}}(\pi^+\varrho^- - \pi^-\varrho^+)$ | $\frac{1}{\sqrt{3}}(\pi^+\varrho^- + \pi^-\varrho^+ - \pi^0\varrho^0)$ |
| −1 | $\frac{1}{\sqrt{2}}(\pi^0\varrho^- - \pi^-\varrho^+)$ | |

An inspection of this table shows immediately that (1) the observation of the decay mode $\varrho^0\pi^0$ implies $I = 0$ for $X$, (2) charged decay modes for $X$ require $I = 1$ [compare the Gell-Mann, Nishijima relation (1.4.3)].

Now consider the amplitude $T_{fi}$ in terms of its isospin components. If we write

$$X \to \pi_1\varrho \to \pi_1\pi_2\pi_3 \tag{A.7.3}$$

then a further application of Table A.7.1 (with $\pi$ replacing $\varrho$) shows that

$$\begin{aligned} T_{fi} &= T_{1,23} + T_{2,31} + T_{3,12} && \text{for } I = 0 \\ T_{fi} &= T_{1,23} - T_{2,31} && \text{for } I = 1 \end{aligned} \tag{A.7.4}$$

where $T_{i,jk}$ implies a change of sign under the exchange of $j$ and $k$.

Next consider the spatial invariance properties of the $T$s. We can represent the decay as

$$\begin{array}{c} X \\ \pi \leftarrow \underset{l,\mathbf{k}}{\cdot} \rightarrow \varrho \underset{L,\mathbf{q}}{\rightarrow} \pi\pi \end{array} \tag{A.7.5}$$

where $l$ and $L$ are orbital angular momenta in the rest frames of $X$ and $\varrho$ respectively, and $\mathbf{k}$ and $\mathbf{q}$ are the relative three momenta in the same frames (compare Fig. 7.7). Then the spin of $X$ is given by $\mathbf{j} = \mathbf{l} + \mathbf{L}$. Since $L = 1$ for $\varrho$-decay the parity of $X$ is given by $(-1)^l$ (the three pions have intrinsic parity $-1$).

Matrix elements for $T$ can then be constructed using the methods of Zemach (1964, 1965). These are based on the fact that a symmetrised traceless tensor of rank $j$ has the same spatial transformation properties as the angular momentum function for a spin $j$. The tensor is constructed for $l$ combinations of the $k$ and $L$ $(= 1)$ of $q$. They are listed below up to $j = 3$, using the lowest possible values of $l$ ($\delta$- and $\varepsilon$-functions are used to reduce the tensor to the appropriate rank)

$j^P = 0^+$ Impossible

$j^P = 0^-$ $L = 1$ $l = 1$,

$$T \sim \delta_{ij} k_i q_j = \mathbf{k} \cdot \mathbf{q}$$

$j^P = 1^+$ $L = 1$ $l = 0{,}2$ use 0 only

$$T \sim \mathbf{q}$$

$j^P = 1^-$ $L = 1$ $l = 1$

$$T \sim \varepsilon_{ijk} k_j q_k = \mathbf{k} \times \mathbf{q}$$

$j^P = 2^+$ $L = 1$ $l = 2$

$$T \sim (\mathbf{k} \times \mathbf{q})_i \, k_j + (\mathbf{k} \times \mathbf{q})_j \, k_i$$

$j^P = 2^-$ $L = 1$ $l = 1{,}2$ use 1 only

$$T \sim k_i q_j + k_j q_i - \tfrac{2}{3} \delta_{ij} \, \mathbf{k} \cdot \mathbf{q}$$

$j^P = 3^+$ $L = 1$ $l = 2{,}4$ use 2 only

$$T \sim k_i k_j q_k - \tfrac{1}{5} \mathbf{k}^2 \, \delta_{ij} q_k - \tfrac{2}{5} \mathbf{k} \cdot \mathbf{q} \, \delta_{ij} k_k$$

$j^P = 3^-$ $L = 1$ $l = 3$

$$T \sim (\mathbf{k} \times \mathbf{q})_i \, (k_j k_k - \tfrac{1}{5} \delta_{jk} \, \mathbf{k}^2)$$

where the $j = 3$ terms must be summed over all permutations.

The above terms must then be linked with the isospin amplitudes (A.7.4). Resonance features for the $\varrho$ may be included by including Breit–Wigner amplitudes in $jk$ in $T_{i,jk}$ or by using an appropriate Veneziano amplitude. The term $|T_{fi}|^2$ must then be taken to yield the Dalitz plot. Obviously the plots which are then obtained can be quite complicated. One simple feature does emerge, however; all matrix elements with $j^P = 1^-, 2^+, 3^-$ contain the term $\mathbf{k} \times \mathbf{q}$. According to the arguments of § 7.1(c) amplitudes containing this term must vanish along the edge of the plot. Thus the observation of a uniform depletion at the edges of a plot leads one to suspect possible spin-parities of $1^-, 2^+, 3^-, \ldots$

In practice the problem is always complicated by background effects, and the assignment of spin-parities is normally given in terms of probabilities and/or confidence levels.

# REFERENCES‡


ALBRECHT, W., BEHREND, H. J., BRASSE, F. W., FLAUGER, W., HULTSCHIG, M., and STEFFEN, K. G. (1967) *Phys. Rev. Lett.* **17**, 1192.

ALFF, C. *et al.* (1962) *Phys. Rev. Lett.* **9**, 325.

ALLABY, J. V. *et al.* (1969) *Phys. Lett.* **30**B, 500.

ANDERSON, C. D. (1932) *Science* **76**, 238.

ASHKIN, J., FAZZINI, T., FIDECARO, G., MERRISON, A. W., PAUL, H., and TOLLESTRUP, A. V. (1959) *Nuovo Cim.* **13**, 1240.

AUGUSTIN *et al.* (1969a) *Lettere al Nuovo Cim.* **2**, 214.

AUGUSTIN *et al.* (1969b) *Phys. Lett.* **28**B, 503.

AUSLANDER, V. L., BULKER, G. I., PESTOV, IU. N., SIDEROV, V. A., SRINSKY, A. N., and KHABAKHPASHEV, A. G. (1967) *Phys. Lett.* **25**B, 433.

BAILEY, J. *et al.* (1968) *Phys. Lett.* **28**B, 287.

BARBARO-GALTIERI, A. *et al.* (1970) *Rev. Mod. Phys.* **42**, 87.

BARGER, V. and CLINE, D. (1966) *Phys. Rev. Lett.* **16**, 913.

BARGER, V. and CLINE, D. (1968) *Phys. Rev. Lett.* **21**, 392.

BARGER, V. and PHILIPPS, R. J. N. (1970) *Phys. Rev. Lett.* **24**, 291.

BARNES, V. E. *et al.* (1964) *Phys. Rev. Lett.* **12**, 204.

BELLETINI, G. (1968) *14th International Conference on High Energy Physics* (Vienna), 338.

BLACKETT, P. M. S. and OCCHIALINI, G. P. S. (1933) *Proc. Roy. Soc.* A **139**, 699.

BLATT, J. M. and WEISSKOPF, V. F. (1952) *Theoretical Nuclear Physics*, Wiley.

BOYD, D. P. *et al.* (1968) *Bull. Am. Phys. Soc.* **13**, 1424.

BURKHARDT, H. (1969) *Dispersion Relation Dynamics*, North Holland.

CABBIBO, N. (1963) *Phys. Rev. Lett.* **10**, 531.

CASSEN, B. and CONDON, E. U. (1936) *Phys. Rev.* **50**, 846.

*CERN Courier* (1970) **10**, 80.

CHAN HONG-MO, RAITIO, R. O., THOMAS, G. H., and TÖRNQVIST, N. A. (1969) CERN preprint Th. 1111.

CHEW, G. F. and FRAUTSCHI, S. (1962) *Phys. Rev. Lett.* **8**, 41.

CHEW, G. F., FRAUTSCHI, S., and MANDELSTAM, S. (1962) *Phys. Rev.* **126**, 1202.

CHRISTENSON, J. H., CRONIN, J. W., FITCH, V. L., and TURLAY, R. (1964) *Phys. Rev. Lett.* **13**, 138.


‡ Reference is given to the first name only in papers with more than six authors.

COHEN, E. R., CROWE, K. M., and DUMOND, J. M. (1957) *Fundamental Constants of Physics*, Interscience.
COLEMAN, S. and GLASHOW, S. L. (1961) *Phys. Rev. Lett.* **6**, 423.
COLLINS, P. D. B., JOHNSON, R. C., and SQUIRES, E. J. (1968) *Phys. Lett.* **27**B, 23.
CONDON, E. U. and SHORTLEY, G. H. (1935) *The Theory of Atomic Spectra*, Cambridge.
COWAN, C. L., REINES, F., HARRISON, F. B., KRUSE, H. W., and MCGUIRE, A. D. (1956) *Science* **124**, 103.

DALITZ, R. H. (1957) *Rep. Progr. Phys.* **20**, 163.
DANBY, G. *et al.* (1962) *Phys. Rev. Lett.* **9**, 36.
DEPOMMIER, P., HEINTZE, J., RUBBIA, C., and SOERGEL, V. (1963) *Phys. Lett.* **5**, 61.
DIEBOLD, R. (1969) SLAC-PUB-673.
DIRAC, P. A. M. (1928) *Proc. Roy. Soc.* A **117**, 610.
DIRAC, P. A. M. (1929) *Proc. Roy. Soc.* A **126**, 360.
DIRAC, P. A. M. (1947) *The Principles of Quantum Mechanics*, Oxford.
DOLEN, R., HORN, D., and SCHMID, C. (1968) *Phys. Rev.* **166**, 1768.
DYSON, F. J. (1949) *Phys. Rev.* **75**, 486, 1736.

D'ESPAGNAT, B. and PRENTKI, J. (1956) *Nuclear Phys.* **1**, 33.

FERMI, E. and YANG, C. N. (1949) *Phys. Rev.* **76**, 1739.
FEYNMAN, R. P. (1949) *Phys. Rev.* **76**, 749, 769.
FILTHUTH, K. (1969) CERN 69-7, 131.
FOLEY, K. J. *et al.* (1967) *Phys. Rev. Lett.* **19**, 193.
FOX, G. C. (1969) *Comments on Nuclear and Elementary Particle Physics* **3**, 19.
FRIEDMAN, J. I. and TELEGDI, V. L. (1957) *Phys. Rev.* **105**, 1681.

GARWIN, R. L., LEDERMAN, L. M., and WEINRICH, M. (1957) *Phys. Rev.* **105**, 1415.
GELL-MANN, M. (1953) *Phys. Rev.* **92**, 833.
GELL-MANN, M. (1956) *Suppl. Nuovo Cim.* **4**, 848.
GELL-MANN, M. (1962) *Phys. Rev.* **125**, 1067.
GELL-MANN, M. (1964) *Phys. Lett.* **8**, 214.
GELL-MANN, M. and PAIS, A. (1955) *Phys. Rev.* **97**, 1387.
GOLDHABER, M., GRODZINS, L., and SUNYAR, A. W. (1958) *Phys. Rev.* **109**, 1015.

HEISENBERG, W. (1932) *Z. für Phys.* **77**, 1.
HEISENBERG, W. (1943) *Z. für Phys.* **120**, 513, 673.
VAN HOVE, L. (1966) *13th International Conference on High Energy Physics* (Berkeley), 253.

IGI, K. (1962) *Phys. Rev. Lett.* **9**, 76.

JAHNKE, E. and EMDE, F. (1945) *Tables of Functions*, Dover.

LEE, T. D. and YANG, C. N. (1956a) *Phys. Rev.* **104**, 254; (1956b) *Nuovo Cim.* **3**, 749.
LEPORE, J. V. (1950) *Phys. Rev.* **79**, 137.
LOGUNOV, A., SOLOVIEV, L. D., and TAVKELIDZE, A. N. (1967) *Phys. Lett.* **24**B, 181.
LOVELACE, C. (1968) *Phys. Lett.* **28**B, 264.

MESHKOV, S., LEVINSON, C. A., and LIPKIN, H. J. (1963) *Phys. Rev. Lett.* **10**, 361.

NISHIJIMA, K. (1955) *Progr. Th. Phys.* **13**, 285.

OKUBO, S. (1962) *Progr. Th. Phys.* **27**, 949.
OLSON, D. N., SCHOPPER, H. F., and WILSON, R. R. (1961) *Phys. Rev. Lett.* **6**, 286.

PAIS, A. (1952) *Phys. Rev.* **86**, 663.
PANOFSKY, W. K. H. (1968) *14th International Conference on High Energy Physics* (Vienna), 23.
PAULI, W. (1933) *Proc. Solvay Conf.*
PEIERLS, R. E. (1954) *Proc. of the Glasgow Conference on Nuclear and Meson Physics*, 296, Pergamon.
*P.E.P.* (1965) *The Physics of Elementary Particles*, Pergamon.
PETERSSON, B. and TÖRNQVIST, N. A. (1969) *Nuclear Phys.* **B 13**, 629.
POIRIER, J. A. and PRIPSTEIN, M. (1963) *Phys. Rev.* **130**, 1171.

REGGE, T. (1959) *Nuovo Cim.* **14**, 951;
REGGE, T. (1960) *Nuovo Cim.* **18**, 947.
REINES, F. and COWAN, C. L. (1959) *Phys. Rev.* **113**, 273.

SCHMID, C. (1968) *Phys. Rev. Lett.* **20**, 689.
SCHOPPER, H. F. (1966) *Weak Interactions and Beta Decay*, North-Holland.
SRIVASTAVA, P. P. and SUDARSHAN, G. (1958) *Phys. Rev.* **110**, 765.
STEINBERGER, J. (1969) CERN 69-7, 291.
DE SWART, J. J. (1963) *Rev. Mod. Phys.* **35**, 916.

TING, S. C. C. (1968) *14th International Conference on High Energy Physics* (Vienna), 45.

VENEZIANO, G. (1968) *Nuovo Cim.* **57**A, 190.

WU, C. S., AMBLER, E., HAYWARD, E. W., HOPPES, D. D., and HUDSON, R. P. (1957) *Phys. Rev.* **105**, 1413.
WU, C. S. and MOSKOWSKI, S. A. (1966) *Beta Decay*, Interscience.

ZEMACH, C. (1964) *Phys. Rev.* **133**B, 1201; (1965) *Phys. Rev.* **140**B, 97, 109.
ZWEIG, G. (1964) CERN preprint 8419/Th. 412.

# INDEX